KU-186-582

HISTORY OF TECHNOLOGY

HISTORY OF TECHNOLOGY

History of Technology

Volume Seventeen, 1995

Edited by
Graham Hollister-Short and
Frank A.J.L. James

MANSELL

First published 1996 by
Mansell Publishing Limited, *A Cassell imprint*
Wellington House, 125 Strand, London WC2R 0BB, England
215 Park Avenue South, New York, New York 10003, USA

British Library Cataloguing-in-Publication Data

History of Technology — 17th volume (1995)
1. Technology — History — Periodicals
609 T15

ISBN 0-7201-2284-8
ISSN 0307-5451

Library of Congress Catalog Card Number: 76-648107

Typeset by York House Typographic Ltd, London
Printed and bound in Great Britain by Biddles Limited, Guildford and King's Lynn

Contents

Editorial

In Volume 16 our two leading papers were concerned with aspects of that most contemporary of all technologies, the know-how required to drill for oil, and then to pump and store it, in hundreds of feet of water in the North Sea, often hundreds of miles offshore. In our present volume, by contrast, the first two papers in the collection, those by Susan Murphy and Dennis Simms, discuss aspects of the technology of the ancient Greek world, namely, the fine engineering requisite for sustained automatic displays as described by Heron of Alexandria (fl. AD 50) and the engineering achievements of Archimedes of Syracuse (d. 212 BC). With respect to the latter, of course, one difficulty is to separate fact from fiction in arriving at any assessment. One thinks, for instance, of such legendary stories as that concerning the burning mirrors. For the rest, the other seven papers in this volume are evenly distributed between medieval and early modern Europe and work on more recent aspects of the history of technology.

Two papers deal with late medieval topics. That by Thomas Glick examines the role of minority groups as agents of technological diffusion. In this case he examines the part played by the Moriscos and the Marranos. The former were the Christianized Arabs expelled from Spain after 1602 in an early episode of what can only be called ethnic cleansing; the Marranos, who suffered likewise, were Jews who had converted to Christianity but who were known to have, or were suspected of having, apostasized. Both took with them into exile the technologies they had practised in Spain. Our own small contribution on medieval cranks refers, of course, not to the users of cranked reading devices but to the rather curious use to which cranks were being put in the fourteenth and fifteenth centuries. This seems to have been happening across Europe, from England in the West to Slovakia in the East. Perhaps even more interesting here is that, although the compound crank is first attested in a drawing of 1335, the precise context in which it occurs, in the *Texaurus* of Guido da Vigevano, really forbids belief in the somewhat chimerical application proposed. If this is the case, then it suggests that the idea of the crank was borrowed from some other real application of the time. The earliest date for a reading device furnished with a compound crank is, on the other hand, only a little later, and might be rather before 1360. It is just within the bounds of possibility, therefore, that Guido da Vigevano, himself a bookman, may have sought to import a bookish piece of technology into the art of war. If necessity is the mother of invention, then the difficulty medieval scholars seem to have experienced in toting heavy tomes was clearly a spur to invention both north and south of the Alps.

David Bridge and Walter Endrei introduce us to the work of sixteenth-century technologists. Among the technological *voyageurs* of the 1560s, carrying with them the latest technological know-how from the Tyrol to England, were the miners from Schwaz, whom Elizabeth I of England had invited to the kingdom to develop the copper deposits of Cumbria. It is their activities in the 1560s which David Bridge relates with clarity and precision. By contrast, that aspect of Jean Errard's work which Walter Endrei has chosen to explore is the technologist lounging in his arm-chair, so to speak. Errard, in his book of machines of 1584, was busily engaged in drafting technological schemes in the pious cause, one supposes, of self-promotion. Whatever may have been Errard's motives, the fact remains that little real work has been done on the machine books in the way of relating them to the contemporary technology to which they are in some sense holding up the mirror; we hope that this paper may stimulate further studies in this field.

Carroll Pursell, in his paper on the adoption of mass-production techniques in the American furniture industry, permits us to pursue further one of the great themes of modern technology: that is, the mechanization of US industry, already ably developed in our previous volume in David Lewis's discussion of automatic pig casting. Mechanization is inextricably woven into the pattern of thought and customs of the United States, and in this study of the furniture industry we are able to see the special forms it took there. In this connection it is interesting to recall the situation in 1948, commented on by Siegfrid Giedion in his *Mechanization Takes Command: A Contribution to Anonymous History*, published in that year. He wrote then of his discovery of an almost complete lack of research in the most basic aspects of how mass society came to be supplied with the 1001 articles it needed (not to mention the 57 varieties under the sign of the gherkin sold in clear glass containers in Pittsburgh). What was even worse was his discovery that 'an amazing historical blindness, [had] prevented the preservation of important historical documents . . . catalogues and advertising leaflets'. Nor perhaps, he went on, 'can one blame the [US] Patent Office for ridding itself (in 1926) of the original patent models. The historians who did not succeed in awakening a feeling for the continuity of history are to blame.'

Last, but not least, we must mention Jennifer Tann's paper on the contribution of diffusion process theory to the history of technology. One of the most important frontiers of the history of technology is the one it shares with the history of ideas. This, in our opinion, should always be regarded as an open frontier, with traffic in both directions to be welcomed as likely to lead to the mutual benefit of both disciplines.

In Memory of Dr Joseph Needham (1901–1995)

I knew Joseph Needham for only the last 15 years of his life, and even then most of our communications were conducted by letter. What I learned from him, and what I owe him, are things difficult to relate without descending into sentimentality, but that is a risk I must take in acknowledging my very considerable debt. He was so very modest and generous in acknowledging the work of those who were able, in even a small way, to assist him in his own great work. This, alas, was something he did not live to see completed. I remember asking him, many years ago, about the scheduled date for completion of *Science and Civilisation in China.* He replied, without a moment's hesitation: 'I shall then be 96 and will go up in a cloud of smoke, which will be most unpleasant for those about me.'

His warmth and generosity are marked for me in a way which any researcher will instantly understand. When one has worked on a paper and finally launched it into the world, there is all too often only a great silence. After I had published a paper in 1979, to my great surprise a packet came through the post containing a pile of offprints and a simple message, signed 'JN', saying: 'Thanks for the sector and chain.' Subsequently, I was able to help him in a very small way as he worked on Volume V, Part 7, 'The Gunpowder Epic'. His help in return whenever I needed it was unfailing. I could always count on what he called a 'trumpet blast' to support any application I might be making. Once, when I needed a photograph, he lent me his only negative.

For the rest, I retain a series of images. At conferences I see him swooping like a hawk between lectures. In his office at Brooklands Avenue, before the move to Robinson College, at tea-time, I remember him smoking green cheroots of a pungency before which even I, as a hardened smoker, quailed. In the same office – this I recall with particular pleasure – I observed all around me the evidence that Joseph ran a horizontal filing system and marked him instantly as 'one of us'.

Above all, as a historian learning how to become a historian of technology, I owed him even more. In reading his luminous pages, written with such elegance and grace, I can only hope that some of this may have rubbed off on my own writing.

Graham Hollister-Short

The Contributors

David Bridge
28 Abbey Vale
St Bees, Cumbria
CA27 0EF
England

Professor Walter Endrei
Eötvös Loránd University
Angyalföldi 24b
H-1134 Budapest
Hungary

Professor Thomas F. Glick
University of Boston
226 Bay State Road
Boston
Massachusetts 02215
USA

Dr Graham Hollister-Short
Department of History of Science
and Technology
Imperial College
London
SW7 2AZ
England

Hellmut Janetschek
Kustos
Technisches Museum-Wien
Wolfganggasse 30
A-1120 Vienna
Austria

Susan Murphy
Department of Classics
Waggener 123
University of Texas at Austin
Austin
Texas 78712
USA

Professor Carroll Pursell
Case Western Reserve University
10900 Euclid Avenue
Cleveland
Ohio 44186-7167
USA

Dr D.L. Simms
17 Dunstan Road
London
NW11 8AG
England

Professor Jennifer Tann
University of Birmingham
Edgbaston
Birmingham
B15 2TT
England

Notes for Contributors

Contributions are welcome and should be sent to the editors. They are considered on the understanding that they are previously unpublished in English and are not on offer to another journal. Papers in French and German will be considered for publication, but an English summary will be required. The editors will also consider publishing English translations of papers already published in languages other than English. Three copies should be submitted, typed in double spacing (including quotations and notes) with a margin, on A4 or American Quarto paper. Include an abstract of 150–200 words and two or three sentences for 'Notes on Contributors'.

It would be appreciated if normal printers' instructions could be used. For example, words to be set in italics should be underlined and *not* put in italics. Authors who have passages originally in Cyrillic or oriental scripts should indicate the system of transliteration they have used. Quotations when long should be inset without quotation marks; when short, in single quotation marks. Spelling should follow the *Oxford English Dictionary*, and arrangement H. Hart, *Rules for Compositors* (Oxford, many editions). Be clear and consistent.

All papers should be rigorously documented, with references to primary and secondary sources typed separately from the text in double spacing and *numbered consecutively*. Cite as follows for books:

> 1. David Gooding, *Experiment and the Making of Meaning: Human Agency in Scientific Observation and Experiment* (Dordrecht, 1990), 54–5.

Subsequent references may be written:

> 3. Gooding, *op. cit.* (1), 43.

Only name the publisher for good reason. For theses, cite University Microfilm order number or at least Dissertations Abstract number. Standard works like DNB, DBB must be thus cited.

And as follows for articles:

> 13. Andrew Nahum, 'The Rotary Aero Engine', *Hist. Tech.*, 1986, 9: 125–66, p. 139.

Line drawings should be drawn boldly in black ink on stout white paper, feint-ruled paper or tracing paper. Photographs should be glossy prints of good contrast and well matched for tonal range. The place of an illustration should be indicated in the margin of the text where it should also be keyed in. Each illustration must be numbered and have a caption. Xerox copies may be sent when the article is first submitted for consideration.

Heron of Alexandria's *On Automaton-Making*

SUSAN MURPHY

ABSTRACT

Heron of Alexandria's *Peri automatopoietikes* (*On Automaton-Making*) is a Hellenistic technical manual describing the construction of two kinds of automata. The first is a mobile shrine to Dionysos with small figurines of the god and dancing bacchant worshippers, which moves on a wheeled base to a specified spot, stops while the figures enact a scene of sacrifice and libation-pouring and then rolls back to its original position. The second automaton is a stationary miniature theatre which stages a complete tragedy by itself when activated. Both automata are powered by the action, on various cords and axles, of a descending counterweight. This paper contains an annotated translation of the *Automatopoietikes* accompanied by an introduction outlining the scholarship on this text and examining its usefulness as a source of information on Hellenistic stagecraft and Hellenistic automata.

INTRODUCTION

Heron of Alexandria's *Peri automatopoietikes* (*On Automaton-Making*)[1] is an exposition of techniques for constructing two kinds of automatic theatrical displays: in Book I (ch. 1–19), a mobile base, which travels to a given point, stops, exhibits a performance by the figures it carries, and withdraws to its original position; in Book II (ch. 20–30), a stationary theatre presents a complete play. Both automata, once set in motion, continue without further assistance until they reach the end of their programme. The mobile base carries a miniature shrine to Dionysos, with a figure of the god inside, a Winged Victory on the roof, and a pair of altars and a ring of dancing Maenads around the outside. The whole assemblage sits on top of four columns which in turn rest on a square, hollow base containing the wheels on which the automaton travels. The author states, in each book, that he is using a specific example in order to explicate a set of techniques which may be generally applied to the building of the type of automaton under discussion.[2] At the same time, it seems that for Heron the attraction of writing about automaton-building is the mechanical complexity involved: 'every facet of mechanics is encompassed in automaton-making, in the completion of its several parts'. The treatise illustrates several basic principles of mechanics, as understood by Heron, and several techniques useful in the building of stationary and mobile automata; it is not, however, a complete how-to manual for constructing either of Heron's automata nor is it a complete textbook on mechanical principles.

Heron wrote in the latter half of the first century after Christ. His date is established by a lunar eclipse of AD 62 described in his *Dioptra*, a set of directions for building and using a theodolite/water level.[3] As well as the treatise on automatic puppet theatres, his works include the *Dioptra*, the *Pneumatika* (a compilation of descriptions of devices powered by steam or water); the *Belopoiika* and *Cheiroballistra* (the latter is thought to be incomplete, but enough has survived for Marsden to reconstruct the machine described[4]), on catapults; the *Katoptrika* on mirrors, the *Definitions* (ὅροι), defining geometrical terms, the *Geometrica*, *Stereometrica* and *On Measures* (περὶ μέτρων); and the *Mechanika*, on mechanical principles. The *Mechanika* survives in a ninth-century Arabic translation and in fragments preserved by Pappos of Alexandria: the other works are extant in Greek. A few of Heron's works are known by title but no longer extant: the *Baroulkos*, describing a set of gears for lifting heavy objects, partly preserved in the *Mechanika* and *Dioptra*; and *On Waterclocks*.[5]

Some doubts have been expressed concerning the originality and validity of Heron's works. H. Diels called him 'ein reiner Banause' and I. Hammer-Jensen doubted that he understood his own writing about pneumatics.[6] On the other hand, A.G. Drachmann, in *Ktesibios, Philon and Heron* (1948), analysed Heron's debt to Philon of Byzantium (*c.* 300 BC), and their mutual and individual debts to Ktesibios (*c.* 370 BC), at least as far as the study of pneumatics is concerned, and concluded that Heron was not only a well-informed student of scientific progress up to his time, making use of Aristotle, Straton and Archimedes, but also a competent engineer, recording the results of his own experience as well as transmitting the work of his predecessors. He advances his own theory that the rough state of much of the *Pneumatika* indicates that it is a collection of notes which was never completely prepared for publication.[7] Although the vagaries of textual transmission are perhaps also responsible for some of the problems, Drachmann's argument has great force, especially in light of his ability to make sense of Heron's text. It is also supported by the *Automatopoietikes*, where Heron shows himself to be a competent mechanic and coherent writer.

The *Pneumatika* begins with a theoretical introduction intended to explain the principles on which the various devices described in the body of the work rely. The *Mechanika*, similarly, discusses mechanical principles as well as their application. E.W. Marsden thought that Heron took two different approaches to writing about artillery in the *Cheiroballistra* and the *Belopoiika*. The *Belopoiika* discusses the development of artillery from the beginning, but leaves off before reaching the technical level of Heron's own time; Marsden places its technical content in the third century BC.[8] The title of the work in one of the manuscripts, *Heronos Ktesibiou Belopoiika*, leads Marsden to suggest that it was Heron's edition of Ktesibios' work on artillery.[9] The *Cheiroballistra*, on the other hand, is a technical description of a torsion arrow-shooter which Marsden thinks was developed in Heron's own time and used by Trajan's army in the Dacian campaigns of AD 101–106. Marsden identifies the artillery-pieces depicted on Trajan's column as Heron's cheiroballistra.[10] Again, Heron displays an interest in the work of his predecessors combined with knowledge of current developments.

When we turn to the *Automatopoietikes*, we find that, in both types of automaton, the motion is controlled by the descent of a lead counterweight through a sort of tube or cylinder. The speed of the counterweight is regulated by the flow of grains of some sort out of a hole in the bottom of the tube. Heron specifies grains of millet or mustard seeds for the mobile automaton and dry sand for the stationary one. In his discussion of Book II, Victor Prou suggested that it is essential to the basic idea of both automata that a single counterweight drives all the motions.[11] This is basically right: though there is a second counterweight in the mobile automaton, it is probably set in motion by the first one, at the appropriate moment. The important thing is that once the counterweight has been set in motion, the automata work all by themselves until they come to the end of their performance. There is no stopping and readjusting things in the middle. The operator opens the lid at the bottom of the tube for the seeds or sand to run out, and stands back as the automaton goes through its paces without further assistance.

In both books of the *Automatopoietikes*, Heron alludes to the existence of earlier treatises on automaton-making. He does not mention a source for his mobile automaton; but at the beginning of his discussion of the stationary one, he cites the now lost treatise of Philon of Byzantium as the best 'and most suitable to didactic purposes'. Although he declares that he wishes to write something new about stationary automata, his remark that, except for one modification and one correction, he does not reject anything Philon wrote on the subject, suggests heavy dependence on Philon. He then remarks that it is most beneficial for readers to see the works of older writers with comparisons or corrections (παραθεωρηθέντα ἢ διορθώσεως) adduced. Heron has just given us a comparison – he prefers to bring Athena on stage by a different method from Philon's – and a correction: Philon left out the explanation of the thunder sound effect, so Heron supplies it. We cannot know, without Philon's treatise on automata, just how closely Heron follows Philon, and whether he gives us his own original version of the Nauplios puppet play, or a revision of Philon's treatise.

The *Automatopoietikes* lacks too many details to be an instruction-manual for constructing either of Heron's automata; but the techniques Heron does describe could, as he says, be applied to other scenarios. He describes the base of a mobile automaton (a box with three wheels – two sharing an axle, one on its own), explains how to make the base move in a straight line and then return to its original position, how to make it trace a circle, a rectangle or a curving path. He describes the mechanism for making the Maenads circle around the shrine to Dionysos, for lighting fires on the altars outside the shrine, for making milk and wine squirt out of the Dionysos figure's cup and staff. He tends to leave out small, but very practical, details, such as exactly how to connect a cord from a given apparatus to the counterweight (see section **12** 3) and how to get all the cords from the axles to the counterweight without getting them tangled up in each other. In section **3**, Heron suggests exact dimensions for the frame of the mobile automaton: the base should be 4 palms by 3 palms by 1 cubit, and the height of the columns on which the Dionysos shrine rests 8 palms; but none of the dimensions of the internal moving parts of the apparatus are given, including the volume (not to mention the exact location) of the tube containing the millet seeds. Similar instances of vagueness occur in Book II, and examples from both books are extensively catalogued in the notes accompanying this translation.

As for mechanical principles, of the five simple machines described in Book II of the *Mechanika* – the wedge, lever, pulley, winch and screw – only the wedge does not appear in the *Automatopoietikes.* The winch, in the form of an axle with a larger drum around it, appears at **26** 8, where the axle driven by the counterweight turns a drum which winds up another cord and thus rotates the pivot around which the cord is originally wound. The screw is used in chapter 10, to raise and lower the alternating sets of wheel which allow the base to turn square corners. At **18** 3 Heron remarks, apropos of larger wheels moving smaller ones, that this is done by means of levers: the ancient writers on mechanics understood leverage in terms of circular motion. In Heron's *Mechanika,* Aristotle's *Mechanical Problems* and Philon's *On Levers,* the arms of a lever are explained as the radii of two concentric but unequal circles. It is relatively easy to move the large circle, and it in turn moves the smaller one.[12] Pulleys appear in various places throughout the treatise, at **5** 5, **13** 7–8, **16** 3, **27** 4, etc. Thus, Heron does make good most of his promise to illustrate all the basic principles of mechanics in the *Automatopoietikes.*

None the less, the *Automatopoietikes* is hardly a satisfactory discussion of mechanical principles; nor is it an exhaustive manual for building automata: as noted above, too many details are missing. Heron perhaps assumes that his readers are familiar with these basic problems, and will know how to solve them. He does intend his treatise to have practical use: he emphasizes the superiority of his devices over those of his predecessors, and claims to be using an up-to-date and impressive scenario for each automaton. It seems, then, that his audience was the craftsmen who specialized in automata, not scientists interested primarily in mechanical principles, but people who might have been interested in the mechanical principles encompassed in their own art, as well in 'up-to-date scenarios'.

Heron's second automaton is a miniature theatre. The context in which it would have functioned is unknown: perhaps it was a sophisticated toy; perhaps it had some connection to full-scale theatre. The first automaton involves a figure of Dionysos, the god particularly connected with Classical Athenian theatre. Athenaeus reports a description, by one of his 'philosophers at dinner', of a procession arranged by Ptolemy Philadelphos which involved a libation-pouring statue of Dionysos, and an 'image (ἄγαλμα) of Nysa, twelve feet high' which could automatically stand, pour a libation of milk and sit down again.[13] He also records the statement that 'the Athenians yielded to Potheinus the marionette-player (*neurospastes*) the very stage on which Euripides and his contemporaries performed their inspired plays.' Aristotle in the *Cosmos*, Xenophon in the *Symposium* and Plato in the *Laws* also mention *neurospasta* – marionette-like puppets controlled by strings, perhaps of animal sinew.[14] The point of the *neurospasta* seems to be lower-class entertainment: Athenaeus and Xenophon in particular refer to them as vulgar. These *neurospasta* are apparently a different class of things from Heron's automata, which seem to be dignified by their complexity and perhaps by their scientific usefulness.

Derek J. De Solla Price, in an article on automata in the history of science, argues that the early automata are evidence of a desire to imitate natural phenomena, which stems from a desire to understand them, and rapidly develops into the impulse to find mechanistic explanations for those phenomena. In other words, he argues that a mechanistic theory of the universe began to develop much earlier than is conventionally thought, and that automata of various sorts, and other mechanical devices, were more often the result than the stimulus of human desire to explain the universe in terms of easily grasped mechanical processes. Price sees the development of various kinds of automata as closely related to the development of scientific theories. He argues from the machines themselves, and from the existence of treatises on automata and mechanics by scientifically minded writers; he does not discuss the more strictly philosophical side of Greek science, nor indeed the unphilosophical motivations for constructing automata.[15]

Aristotle, in the *Generation of Animals*, uses an analogy with 'miraculous automata' (αὐτόματα τῶν θαυμάτων) in explaining the causes of embryonic development – both automata and embryos, once set in motion, continue to their preordained conclusion, without any further contact.[16] In the *Cosmos*, Aristotle uses the ability to set an automaton (here *neurospaston*) in motion by a single act as an analogy for divine influence on the universe.[17] The Xenophon reference (above, n. 14) suggests that puppet shows were popular with people at a lower social level than Xenophon himself, and that their object was entertainment.[18] This kind of puppet differs from Heron's automata, but the passage presents a counterpoint to De Solla Price's view that mechanisms which imitate life were the teething-rings of theoretical science. Other instances of automata, such as the figures of Mars and Venus reported by Claudian (if genuine), or the automatic temple doors and holy water vending machine in Heron's *Pneumatika*, possessed a religious character of some sort, whether they were intended to be snake oil or sacramentals.[19] On the one hand, Heron and his ilk were interested in mechanical principles as well as mechanical devices, and in other branches of science as well;

but others simply wanted toys, entertainment or impressive devices for their temples, and such people were not necessarily interested in philosophy or science.

The second book of the *Automatopoietikes* outlines an automatic version of the story of Nauplios and the shipwreck of Ajax, complete with an appearance by Athena. This could have been based on a staged version, such as the *Nauplios Pyrkaeus* of Sophocles.[20] It might seem at first glance that a puppet theatre is unlikely to have provoked a reaction other than amusement; but the success of Peter Arnott at staging serious drama, including Greek tragedy, with puppets, suggests that this is not necessarily the case.[21] However, Arnott's puppet plays were actual adaptations of dramas intended for the stage, with Arnott supplying the voices. If Heron intended that any voices or sound effects accompany his Nauplios drama, beside those mechanically produced, he does not mention them. Because of the sketchiness of the detail he presents (How much cord? How much sand?) it is not clear how long the production was supposed to take. If the purpose of such a display was not simply to entertain someone's guests, perhaps it was a short, silent prelude to a full-scale drama: the cartoon before the feature.

J. Formigé suggests that Heron's second automaton may be a source for large-scale Hellenistic stagecraft.[22] The staging devices apparent in the stationary puppet theatre are: opening and closing doors; objects (moulded arms attached to figures painted on the wall) moved by spiked wheels striking levers behind the wall to which the objects are attached; objects (dolphins) which rise from a space under the stage through the floor and dive down again, by means of spokes attached to revolving axles; a painted roll of papyrus which is scrolled from one axis to another on the opposite side of the stage, to give an effect like a moving picture; painted drop-cloths unrolling from the ceiling; fires which blaze up from concealed lamps when their gratings are opened; a figure (of Athena) which is pulled up from some hiding-place, pulled around in a circle and hidden again; lead balls bouncing off a taut piece of leather to create a peal of thunder; and a painted piece of scenery which is dropped down into the theatre and through the floor from above with the aid of wires to guide it through the slots in floor and ceiling. Two of these devices are conspicuously compared or contrasted with their counterparts in large theatres. The canister of lead balls used for thunder is described as being the same device used in full-scale theatre; and the device for moving Athena is offered as an alternative to a small-scale imitation of the famous crane (*mechané*) used in the Greek theatre since the fifth century. The others are not so obviously related to the theatre, but some connection may exist.

The later author, Pollux (second century AD), gives a list of stage devices current in his own period, whose validity for earlier periods is uncertain.[23] However, his list includes three – the *ekkyklema*, the platform which could be rolled out from backstage to show an interior scene, the *mechané* and the *periaktoi*, revolving prism-shaped units whose three facets could be painted to indicate different settings – which are known to have functioned in Classical Greek stagecraft. The *bronteion*, the thunder canister, is among the stage devices listed by Pollux, but unattested outside that author, at least for the Classical period.[24] Heron supports it for the Hellenistic period. Pollux also mentions a *keraunoskopeion*, which Walton describes as 'a lightning machine that may have worked in a number of unlikely ways, one of which involved a revolving *periaktos*'.[25] Heron's way of producing a lightning bolt on stage is to have it painted on the board that falls through the slots, which seems elaborately dangerous for a full-scale theatre. A pair of falling drop-cloths in succession might have worked. For the Hellenistic and Roman theatre, some or all the rest of Pollux's devices may have existed. 'Charon's stairs', for instance, a stairway under a trapdoor for removing, for example, a ghost from the scene, are still visible, according to J. Formigé, at Arles and Vaison.[26]

Formigé suggests that the rotation of the *periaktoi* may have been synchronized in the same way as the doors in the stationary puppet theatre, by means of a horizontal axle whose rotation drives the vertical axles of the doorposts. He thinks that the use of ropes or cables wound around the axles would have been more difficult for the *periaktoi* than it was for the much smaller puppet theatre doors, and proposes the use of toothed wheels 'such as those described by Vitruvius' instead. The reference is probably to the toothed wheels of Vitruvius **10** 5 – one horizontal, one vertical – for transmitting the motion of a water-wheel to a millstone. Formigé goes on to say that there are many round holes left in the remains of stages at Arles, Dougga, Fréjus, Pompeii, Vaison and Vintimille which could have been the anchors for 'pivots moteurs'.[27]

Formigé also considers the translational motion of the mobile automaton along rails (see *Automatopoietikes* **2** 2) to be a carry-over from the theatre. Mobile scenery pieces, the *scena ductilis*, were slid into place along rails.[28] Overall, however, it is the principle of motion controlled by the counterweight which seems likely to have been the most important connection between puppet stagecraft and its full-scale counterpart. Formigé and Prou emphasize the importance of this feature, especially in comparison with modern but pre-electrical stagecraft.[29] Formigé asserts that the same predominance of counterweights and pulleys appears in the Paris Opéra, before the advent of electricity, as in Heron of Alexandria's automata, and that such techniques must indeed go back to the ancient Greeks: 'on trouve les mêmes machines au théâtre de Syracuse et à l'Opéra de Paris'.[30]

The text of *On Automaton-Making* is preserved in several manuscripts, principal among which seem to be the four primarily used by W. Schmidt in the Teubner edition (Leipzig, 1899, reprinted Stuttgart 1976): codex Marcianus 516, of the thirteenth century; cod. Gudianus 19, sixteenth century; cod. Taurinensis B, V, 20, of 1541; cod. Magliabecchianus II.III 36, sixteenth century.[31] Thévenot used a group of manuscripts in Paris for his *editio princeps* of 1693,[32] as did V. Prou in his study of the treatise. The earliest of the Parisian manuscripts, P.2428, dates to the fifteenth century, the others to the sixteenth; Prou believed that they, and the Vatican manuscript used by Baldassare Baldi, were copied from the same source.[33] Prou, a civil engineer and classicist, published his study of the text, *Les Théâtres d'automates en Grèce au IIᵉ siècle*, in 1881 in the *Mémoires présentés par divers savants à l'Académie des inscriptions et belles-lettres* (ser.I, vol.IX, pt II, Paris, 117ff). It was republished in 1884 as a book, and it is to this edition that I refer in this paper. As well as illuminating many of the practical aspects of the treatise, Prou included his own edition, with running translation, of the second part, on stationary automata. Baldi had published his Italian translation in 1589, just over a century before Thévenot's edition, which uses Baldi's illustrations. Thévenot's edition itself included a Latin translation by a scholar named Couture. Another Latin translation, by Joseph d'Auria of Naples, who also translated other Greek mathematicians into Latin, is in the Bibliothèque Nationale in Paris.[34] The Teubner edition contains a German translation by Schmidt. To the best of my knowledge, mine is the first translation of this text into English.

For the most part, I follow W. Schmidt's Teubner text, with some reference to Prou's text in the second part. I have followed some of Schmidt's emendations for some of the more difficult or corrupt passages, while ignoring some of his lacunas, when the text was translatable without emendation. Translation, rather than textual criticism, has been my main objective: preparing a new edition of Heron's *Automatopoietikes* would have been beyond the scope of the present project. Occasionally, however, I have been forced to adopt some variant reading or disagree with the editor's assessment of a passage. Where I have disagreed with Schmidt or followed alternate readings proposed in his apparatus, I have so indicated in the notes.

The manuscripts contain illustrations, which may ultimately have come from Heron himself; however, they are said to have been badly distorted in transmission; I have not seen them myself. Schmidt's edition is illustrated by H. Querfurth, who redrew the existing figures and supplied new ones of his own, following Schmidt's translation. In some instances Querfurth's illustrations are illuminating; in others, a bit mystifying. Thévenot's edition is illustrated with, according to Prou, Baldi's illustrations. Prou supplies a few illustrations of his own. My own illustrations (like everyone else's) are attempts to reconstruct what is described in the text: if I have added anything not mentioned in the text, I have so indicated in the notes or in the illustration itself.[35]

Notes and References

1. The title of the work given, in the Paris manuscripts, as περὶ αὐτοματοποιητικῶν, and in two of Schmidt's as περὶ αὐτοματοποιητικῆς. Neither title is consistent with the forms of other titles of technical works, such as Heron's *Belopoiika*. Prou thinks that αὐτοματοποιικά would be better. Tittel, in Pauly-Wissowa, agrees that the neuter (-α), rather than the feminine (-ῆς), is more common for titles of technical treatises, and that the genitive plural (-ῶν) is also unusual enough to be suspect. In some of the manuscripts, the second part has a subtitle, περὶ στατῶν αὐτοματῶν, which Schmidt and Prou use, but Thévenot does not. In this paper I have followed Schmidt's use of titles for the sake of convenience, as my translation is based on his text. (See Bibliography, p. 10, for details of these works.)

2. See sections **1** 8 and **21** 1–2.

3. A.G. Drachmann, *The Mechanical Technology of Greek and Roman Antiquity: A Study of the Literary Sources: Acta Historica Scientiarum Naturalium et Medicinalium edidit Bibliotheca Univ. Hauniensis*, vol. 17, (3–140), (Copenhagen, 1963), 9–10, 12, 21, esp. 12, for date and partial list of works. Drachmann follows O. Neugebauer, 'Über ein Methode zur Distanzbestimmung Alexandria-Rom bei Heron', *Kgl. Danske Vidensk. Selsk. Hist.-filol. Medelelser*, 1938, **26** 2.

4. E.W. Marsden, *Greek and Roman Artillery: Technical Treatises* (Oxford, 1971), pp. 2–3, 206–33.

5. *Oxford Classical Dictionary* sv. 'Heron of Alexandria'; I have followed Drachmann in using Greek spellings for the titles, except for those the *OCD* gives in English.

6. References in Drachmann, *op. cit.* (3), 77 and 105.

7. Drachmann, *op. cit.*, (3), 105, 88.

8. E.W. Marsden, *Greek and Roman Artillery: Historical Development* (Oxford, 1969), 3.

9. Marsden, *op. cit.* (4), 2.

10. Marsden, *op. cit.* (8), 4, 189.

11. Victor Prou, *Les Théâtres d'automates en Grèce au IIe siècle* (Paris, 1884), 47– 53.

12. Drachmann, *op. cit.* (3), 61–3, for Heron's *Mech.* **2** 7–8 in English; also 50–1; Schmidt, 399, n. 3, Aristotle, *Mechanical Problems*, 850 a35– b11; this translation, n. 40.

13. *Deipnosophistae*, 5.198c–f (trans. from Loeb ed.); Prou, *op. cit.* (11), 6–7. The statue of Dionysos seems to have simply been a statue in a libation-pouring pose, possibly with some internal piping so as actually to be spouting wine (cf. *Aut.* ch.13); the Nysa is clearly an automaton.

14. Plato, *Laws* 644D; Xenophon, Symposium 4.55. At *Laws* 804B and *Republic* 514, Plato mentions θαυματα and θαυματοποιοι, 'miraculous' puppets and puppeteers. Cf. Prou, *op. cit.* (11), 7, n. 7.

15. De Solla Price, Derek J., 'Automata and the Origins of Mechanism and Mechanistic Philosophy', *Technology and Culture* (1964), 5.

16. *GA*, 734b11, 13–19 (Loeb); ὧν τὸ πρῶτον ὅταν τι κινήσῃ τῶν ἔξωθεν, εὐθὺς τὸ ἐχόμενον γίγνεται ἐνεργείᾳ ... ὥσπερ οὖν ἐν τοῖς αὐτομάτοις, τρόπον μέν τινα ἐκεῖνο κινεῖ οὐχ ἁπτόμενον νῦν οὐθενός, ἁψάμενον μέντοι, ὁμοίως δὲ καὶ τὸ ἀφ' οὗ τὸ σπέρμα ἢ τὸ ποιῆσαν τὸ σπέρμα, αψάμενον μέν τινος, οὐχ ἁπτόμενον δ' ἔτι· τρόπον δέ τινα ἡ ἐνοῦσα κίνησις, ὥσπερ ἡ οἰκοδόμησις τήν οἰκίαν. Cf.741b9.

17. ἀλλὰ τοῦτο ἦν τὸ θειότατον τὸ μετὰ ῥᾳστώνης καὶ ἁπλῆς κινήσεως παντοδαπὰς ἀποτελεῖν ἰδέας, ὥσπερ ἀμέλει δρῶσιν οἱ μηχανοποιοί, διὰ μιᾶς ὀργάνου σχαστηρίας πολλὰς καὶ ποικίλας ἐνεργείας ἀποτελοῦντες. ὁμοίως δὲ καὶ οἱ νευροσπάσται μίαν μήρινθον ἐπισπασάμενοι ποιοῦσι καὶ αὐχένα κινεῖσθαι καί χεῖρα τοῦ ζῴου καὶ ὦμον καὶ ὀφθαλμόν, ἔστι δὲ ὅτε πάντα τὰ μέρη, μετά τινος εὐρυθμίας. *On the Cosmos*, 6.398b13–20, Loeb trans., D.J. Furley: 'the most divine thing of all is to produce all kinds of results easily by means of a single motion, just like the operators of machines, who produce many varied activities by means of the machine's single release-mechanism. In the same way too the men who run puppet shows, by pulling a single string, make the creature's neck move, and his hand and shoulder and eye, and sometimes every part of his body, according to a rhythmical pattern.'

18. *Symposium* 4.55, cf. Prou, *op. cit.* (11), 7, n. 7.

19. *Pneum.*, 1.38, 39, temple doors; 1.21 holy water. Aristotle also mentions temple-door automata, *Mech.* 848a20–38. Claudian, *Carm. Min.* 29 Loeb ed. (48 Gesner), tells of a shrine containing an iron statue of Mars and a lodestone one of Venus. The priest celebrates the marriage ritual, and the lodestone pulls the iron to it. There are no incidental corroborative details, such as where this shrine was located. See also Benjamin Farrington, *Greek Science*, (Harmondsworth, 1961), 200.

20. Prou, *op. cit.* (11), 42; Weil, p. 417; Pauly-Wissowa, Vol. 8, column 1051.

21. Peter Arnott, *Plays without People* (Bloomington, Ind., 1964), *passim.* I have been told by eyewitnesses that Arnott's staging of the *Medea* was very effective, and that his *Bacchae* was 'the scariest thing I've ever seen'.

22. Jules Formigé, 'Note sur les machines des décors mobiles dans les théâtres antiques', *BSAF* (1921), 190–5.

23. *Onomasticon*, IV.123–32; cf. Margaret Bieber, *The History of the Greek and Roman Theater* (Princeton, NJ, 1961), 74.

24. Michael Walton, *Greek Theatre Practice* (London, 1991), 94.
25. *Ibid.*, 98.
26. Formigé, *op. cit.* (22), 194–5.
27. *Ibid.*, 194.
28. *Ibid.*, 192–3. Cf. Bieber, pp. 74–5; Vitruvius 5.6.8.
29. Prou, *op. cit.* (11), 47–53.
30. *Ibid.*, 191.
31. See Schmidt, *op. cit.* (1), 53–6, for a more thorough discussion of the manuscript tradition.
32. Thévenot, ed., *Veterum mathematicorum opera* (Paris, 1693), 243–74.
33. Prou, *op. cit.* (11), 19–21.
34. *Ibid.*, 9–10. Baldi, *Di Herone alessandrino degli automati* (Venice, 1589; New York, Readex Microprint, 1973).
35. Most of my drawings have been made using Claris CAD 2.0 v.3 for Macintosh. I thank Martin Harriman and Deborah Cardillo for allowing me to use their copy of this software.

Bibliography

Peter Arnott, *Plays without People* (Bloomington, IN, 1964).

William Beare, *The Roman Stage*, 3rd edn. (Northampton, 1964), pp. 196–218.

Margaret Bieber, *The History of the Greek and Roman Theater* (Princeton, NJ, 1961), pp. 74–9.

Derek J. De Solla Price, 'Automata and the Origins of Mechanism and Mechanistic Philosophy', *Technology and Culture*, 1964, 5: 9–23.

A.G. Drachmann, 'The Mechanical Technology of Greek and Roman Antiquity: A Study of the Literary Sources', *Acta Historica Scientiarum Naturalium et Medicinalium edidit Bibliotheca Univ. Hauniensis*, Vol. 17 (Copenhagen, 1963), pp. 3–140.

A.G. Drachmann, 'Ktesibios, Philon, and Heron: A Study in Ancient Pneumatics', *Acta Historica Scientiarum Naturalium et Medicinalium edidit Bibliotheca Univ. Hauniensis*, Vol. 4 (Copenhagen, 1948), pp. 74ff.

Benjamin Farrington, *Greek Science: Its Meaning for Us* (Harmondsworth, 1961), pp. 197–200.

Jules Formigé, 'Note sur les machines des décors mobiles dans les théâtres antiques', *BSAF*, 1921, pp. 190–5.

W. Schmidt (ed.), Heron Alexandrinus, Περὶ Αὐτοματοποιητικῆ, *Opera*, Vol. 1, pp. 338–452 (Leipzig, 1899; reprint Stuttgart, 1976).

E.W. Marsden, *Greek and Roman Artillery: Historical Development* (Oxford, 1969).

E.W. Marsden, *Greek and Roman Artillery: Technical Treatises* (Oxford, 1971), pp. 2–3, 206–33.

Pauly-Wissowa, *Paulys Real-Encyclopädie der classische Altertumswissenschaft*, ed. A. F. von Pauly and Georg Wissowa (Stuttgart, 1894–1919).

Victor Prou, *Les théâtres d'automates en Grèce au IIe siècle avant l'ère chretienne d'après les Αὐτοματοποιητικα d'Héron d'Alexandrie: extrait des memoires présentés par divers savants à l'Academie des Inscriptions et Belles-lettres* (Ire série, t. IX, IIE partie) (Paris, 1881).

Thévenot (ed.), trans. Couture, 'Ἥρωνος Ἀλεξανδρέως περὶ αὐτοματοποιητικῶν/ Heronis Alexandrini de automatorum fabrica', *Veterum Mathematicorum Opera*, 243–74.
J. Michael Walton, *Greek Theatre Practice* (London, 1991), pp. 83–108.
Henri Weil, 'Les théâtres d'automates en Grèce ... par Victor Prou, ingénieur civil. (Extrait des Mémoires présentés, Ire ser., t. IX, IIe partie), Paris, 1881', *Journal des Savants*, July 1882, pp. 416–24.

ON AUTOMATON-MAKING

Book I

1 1. The study of automaton-making has been considered by our predecessors worthy of acceptance, both because of the complexity of the craftsmanship involved and because of the striking nature of the spectacle. For, to speak briefly, every facet of mechanics is encompassed within automaton-making, in the completion of its several parts. 2. These are the topics to be discussed: shrines or altars of appropriate size are constructed, which move forward of themselves and stop at specified locations; and each of the figures inside them moves independently according to the argument of the arrangement or story; and then they move back to their original position. Thus such realizations of automata are called mobile. 3. But there is another kind, which is called stationary, and its function is as follows: a toy stage with open doors stands on a pillar, and inside it an arrangement of figures has been set up in line with some story. 4. To begin with, the stage is closed, and then the doors open by themselves, and the painted representation of the figures is displayed. After a little while the doors close and open again of their own accord, and another arrangement of figures, sequential to the first one, appears. Again the doors are closed and opened and yet another arrangement, which logically follows the one before it, appears; and either this completes the planned story, or yet another display appears after this one, until the story finally is finished. 5. And when the figures which have been described are shown in the theatre each one can be shown in motion, if the story demands; for instance, some sawing, some chopping with the adze, some working with hammers or axes – making a noise with each blow, just as they would in real life. 6. Other movements can be effected below the stage; for instance, lighting fires or making figures which were not visible at first appear and then disappear again. Simply, anyone can move the figures as he chooses, without anybody being near them. 7. But the mechanism of the stationary automata is safer and less risky and more adaptable to every requirement than that of the moving ones. Older generations called such feats of craftsmanship miraculous because they offered an amazing spectacle. 8. Therefore, in this book I am going to write about moving automata, and set out my own complex scenario, which is adaptable to every other scenario, so that someone who wanted

to offer a different presentation would not lack anything for the implementation of his own scenario. In the following book I talk about stationary automata.

2 1. First you need a hard, level, smooth surface on which the automaton will move, so that the wheels will neither sink because they are weighed down by their load, nor drag on account of the roughness, nor stop and roll backwards when they hit a bump. 2. If such a surface as has been posited does not exist, you must put straightened slats on the floor, on which there will be grooves formed by rods nailed lengthwise to make the wheels run in the grooves.[1] You must make the mobile automata of light, dry wood, and if there is anything other than wood in the construction, it too should be as light as possible to prevent the mechanism from being hindered by its own weight.[2] 3. And anything that turns or moves in a circle must be perfectly round; and anything that it revolves around must be smooth, not rough, e.g. the wheels around iron pivots inserted into iron sockets, and the figures mounted on bronze axles inserted into bronze axle-boxes made perfectly flush and airtight with them. 4. You must oil these fixtures well so that they will revolve freely with no jamming. Otherwise none of the aforementioned events will happen according to plan. The cords which we use on these moving parts must not stretch or shrink, but remain the length they were at the beginning. 5. We achieve this by stretching the cords tightly between pegs, leaving them for a while, then stretching them again; and after frequent repetitions of this operation, we smear them with a mixture of wax and resin. It is better to put a weight on them and leave them for a good while. When cord is pre-stretched in this way, it will not stretch further, or at most very little. If we find, after setting up the automaton, that one of the cords has stretched, we must trim it again.[3] 6. But you must not use sinew, except perhaps when you need to use a springboard, since this material stretches and shrinks in response to atmospheric conditions. The springboard should be analogous to the arm set in the half-skein of a catapult, as will be made clear at the proper point. All these mobile automata are set in motion either by a springboard or by lead counterweights.[4] 7. Between the source of motion and the part being moved is a cord with one end attached to each. The axle around which the cord is wound is what is moved. Wheels are fixed on the axle so that when the axle is turned and the cord unwound, the wheels, which rest on the floor, are also turned. The base of the moving automaton camouflages the wheels. 8. You must adjust the tension of the springboard or the heaviness of the counterweight so that neither the weight nor the tension is outweighed by the base.[5] Movements in a forward direction result from the action of all the cords, which are looped around the moving parts and attached to the counterweight. The counterweight is fitted into a tube in which it is able to move up and down easily and precisely. 9. For moving automata, millet or mustard seeds are poured into the tube because they are both smooth and slippery, whereas the stationary ones use dry sand. When these grains funnel out from the bottom of the tube, the counterweight sinks gently and causes movement by pulling on each cord. The motion originates from the tension of the cord and ends with the release of tension when the loop slips off the peg on the object being moved. 10. Although the cords are all drawn at an equal rate by the counterweight, the actual movements vary in speed because the cords are wound around objects of different diameter, some larger, some smaller. But cords belonging to objects which move at different times must not be tightened simultaneously: those connected to things which move later should have some slack to them.[6] 11. You must form kinks out of the slack and stick them down with wax at the appropriate places inside the base, so that the counterweight straightens the string gently as it takes up the slack. You must make sure each cord is looped properly around its own axle,[7] and not wound incorrectly, because if one of them is switched or wound incorrectly, the whole thing will come to a standstill. 12. You must avoid old-fashioned scenarios so that your presentation will look modern; for it is possible, as I said before, to create different and varied scenarios while still using the same methods. Your scenario will turn out better if it is well designed. The scenario I am describing is one such.[8]

3 1. Let there be a base one cubit long, four palms wide, three palms high, with a groove running around its upper and lower parts. Set four little columns at its corners, each eight palms high and two palms wide, with some base-mouldings at the bottom and matching capitals on top. Upon the capitals rests a sort of circular architrave one-eighth the height of the whole column, i.e. five fingers.[9] 2. On top of the architrave are boards covering its upper surface, and a groove around it. On the middle of the covering stands a round, free-standing, little shrine with six columns. On this stands a little cone-shaped projection with a raised surface, as stated.[10] 3. A Nike, with wings spread and a wreath in her right hand, stands on the apex. A figure of Dionysos stands in the middle of the shrine holding a thyrsus in his left hand and a chalice in his right. A miniature panther sits at Dionysos' feet. 4. At the position on the platform directly behind and directly in front of the Dionysos is an altar stacked with dry, easily ignitable wood shavings made from planing the boards. Outside Dionysos' shrine, in line with each pillar, are Maenads in whatever pose you prefer.

4 1. These preparations done, when the automaton is set down somewhere and we stand back, after a little while it will move to some prearranged spot. When it stops, the altar in front of the Dionysos will blaze up; and either milk or water will squirt from Dionysos' thyrsus, while wine is poured from his chalice on to the little panther lying beneath it. 2. Each side facing the base's four columns will be decked with garlands. The Maenads will dance in a circle around the shrine, and there will be a clamour of drums and cymbals. Then, when the noise has subsided, the figure of Dionysos will rotate towards the outside and, at the same time, the Nike standing on the rooftop will rotate with him. 3. And when the altar that started behind Dionysos arrives in front of him, it will blaze up again; and again the thyrsus will squirt milk and the chalice pour

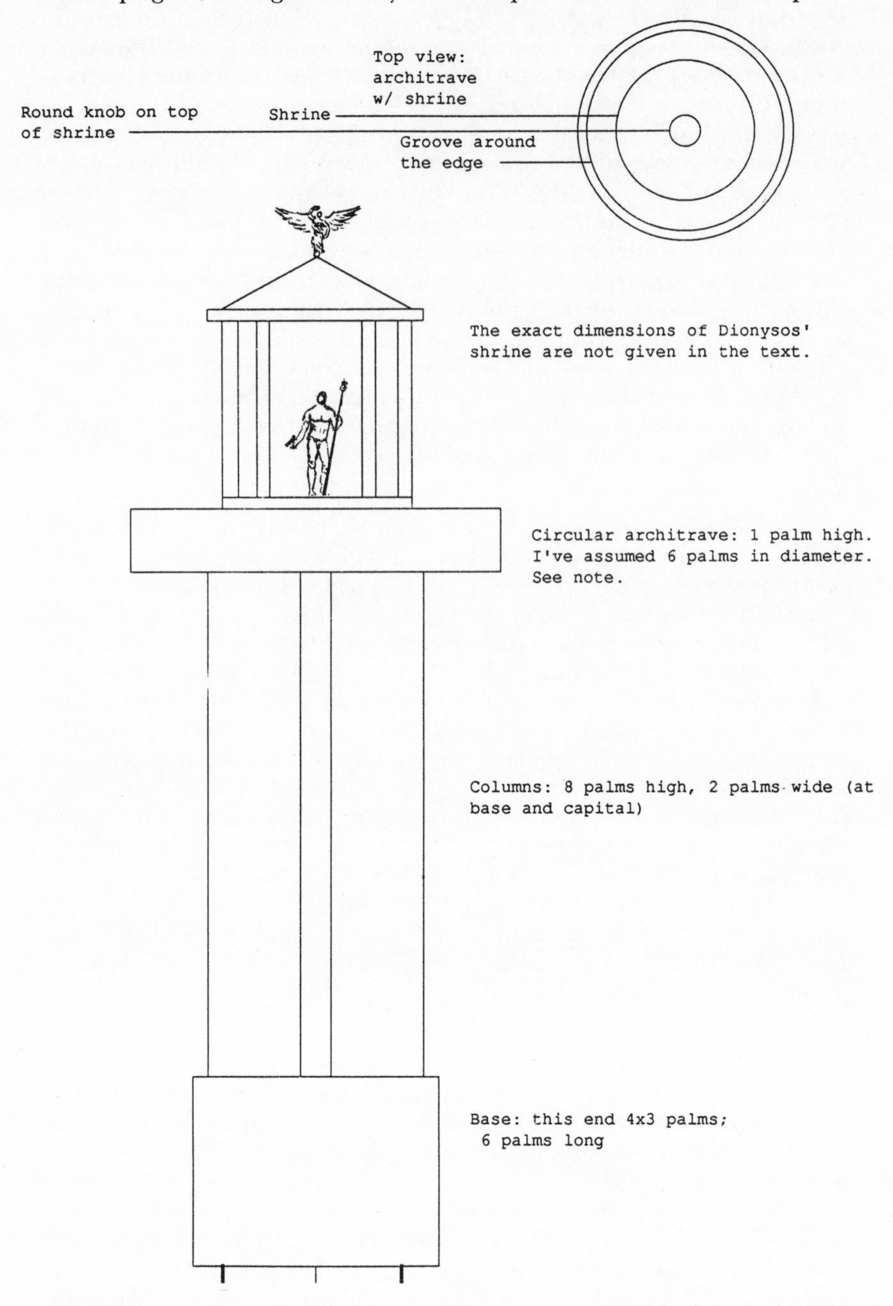

Figure 1

wine. The Maenads will again dance around the shrine to the accompaniment of drums and cymbals; and when they stop the automaton will return to the place from which it started. 4. This is how the display reaches its conclusion. We had to employ the measurements previously mentioned, for the spectacle, were it any bigger, would arouse the suspicion that someone was working these effects from inside. Therefore, in both the moving and the stationary automata, you must be careful of size because of the resultant scepticism.[11] Now that the scenario has been described, I shall go on to the construction of each of its parts.

5 1. Our predecessors have shown us a method of achieving movement and return on the same track, which is difficult and risky, because it rarely happens that someone who follows the instructions they have written down succeeds – as is obvious to whoever has tried them. 2. I, on the other hand, shall show how motion and return along a straight line may be easily and safely achieved, and furthermore, how a base or a figure may be transported around a given circle or even around a rectangle. 3. First, we shall discuss the problem of the straight line. Let there be a base ΑΒΓΔ and let an axle ΕΖ be set in it, turning on pins and anchored to sockets in the wall of the base. Let two wheels, of equal size, with bevelled edges, ΗΘ and ΚΛ, be attached to this axle; and let there be a cylinder ΜΝ similarly attached to the axle, and centred on it. The cord will be wound around this cylinder. 4. Let there be a knob, Ξ, projecting from the cylinder for the string to be looped around. Let there be another wheel, ΟΠ, in the middle of side ΓΔ, revolving in a frame ΡΣΤΥ, around a very small axle, ΦΧ. Let the axles of the wheels be so adjusted that the base stands level at all points. After the loop of the cord is hooked around the knob Ξ, let the string be wound around the cylinder. 5. Then, after setting a rectangular tube on end in the middle of the base, let the other end of the cord be threaded through a sheave at the top of the tube and tied to the lead counterweight inside it. Then if anything causes the weight to sink down in the tube, it pulls on the cord, which, as it is unwound from the cylinder, turns the wheels ΗΘ and ΚΛ; and they in turn, as they roll along the smooth surface, will move the base until the loop falls off the peg or the weight comes to rest on something.[12]

6 1. The movement is effected in the way just described, and the return as follows: when the cord has been partially wound around the cylinder, let the other end be looped around knob Ξ and wound around the cylinder in the opposite direction from the first part. Then let it be similarly attached to the counterweight with a ring to hold it firm. When the counterweight sinks, it will unwind the first coil and the base will be set in motion. 2. Then, after the cord drops off the knob, it will turn the wheels the opposite way, and thus the return of the base will be effected. But if we want the base to stand for a time after its initial movement, and then return in this way, we wind the cord and loop it around the knob, but instead of immediately winding it the opposite way, take stretches of it, glue them on to the cylinder and then wind the [remaining] cord in the

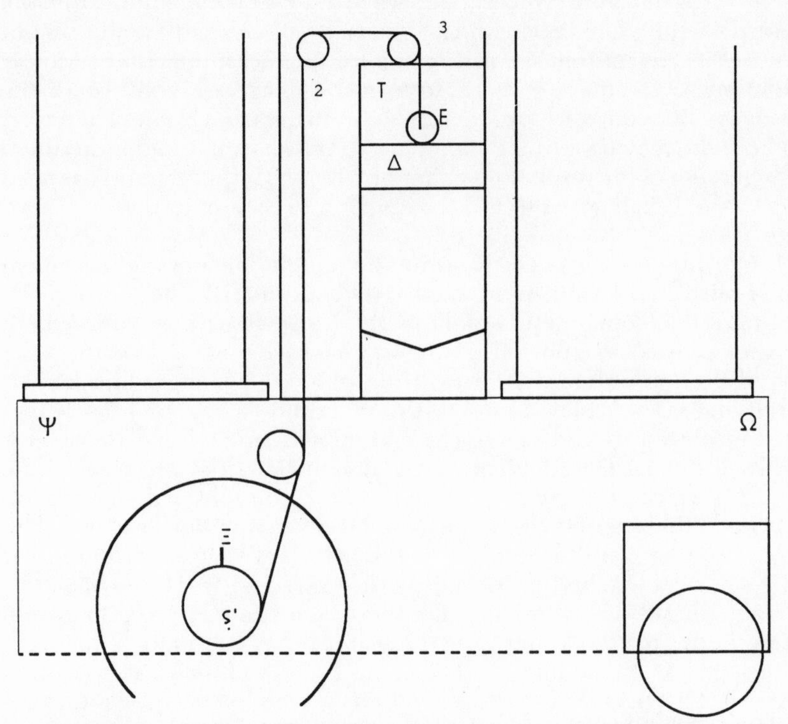

A. Arrangement of cord and counterweight. (Pulleys 1, 2 and 3 are not mentioned in the text.)

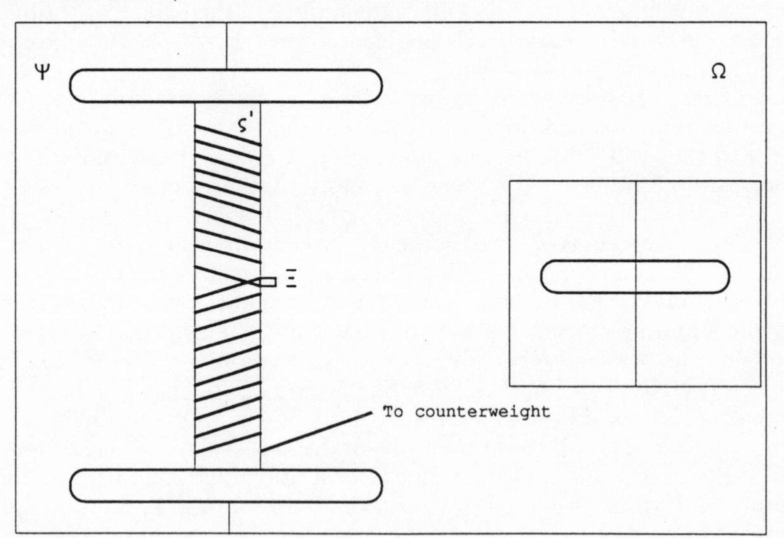

B. Wheels and axles for straight-line motion and return (Book 1, chapters 5–6).

Figure 2

opposite direction and attach it to the counterweight. Then what was described above will happen. 3. But if we also want the base to move back and forth often, we must make more frequent cross-windings with intervals between them of whatever size we choose: we must set the timings for the deities according to preference, by means of lengths of cord.[13] 4. Imagine the base, with the tube, as seen from the side, and let the base be ΨΩ, the cylinder ς', the tube T, the cord AB set around the sheave φ, the counterweight Δ and the ring on it, E.

7 1. A circular motion is achieved as follows. Let the circle around which the base must be conveyed be ABΓ, and its centre, Δ. And let a radius AΔ be measured off, and let EAZ be perpendicular to it at A. Let EZ be the diameter of one of the three wheels, of whatever size we may choose. Let the midpoint of this line be at A, and let ΔE and ΔZ be joined at the top [at Δ]. 2. Let AH be equal to the size of the axle of the wheels, and let HΘK be parallel to EZ. But let the base be MΛNΞ; and let NΞ be parallel to AΔ. And let there be another line, ΔO, which ΠP bisects at right angles below O. The wheels will be oriented with their diameters along EZ, ΘK, ΠP, but their axles will be at TY and OX. 3. Thus the wheels will be so oriented that the base, set on top of them, is evenly balanced. The axle-pins will be at the points T, Y, O, X. Let the cylinder round which the cord is being wound again be placed between A and H, and let things proceed as stated above. Thus the base will be conveyed around the aforementioned circle.[14]

8 1. If a cone is rolled on a level surface its base will describe a circle whose distance from its centre is equal to the side of the cone; but its vertex will remain motionless, being the centre of the said circle. The wheels EZ, ΘK and ΠP are in two cones, whose bases are the circles EZ and ΠP, and

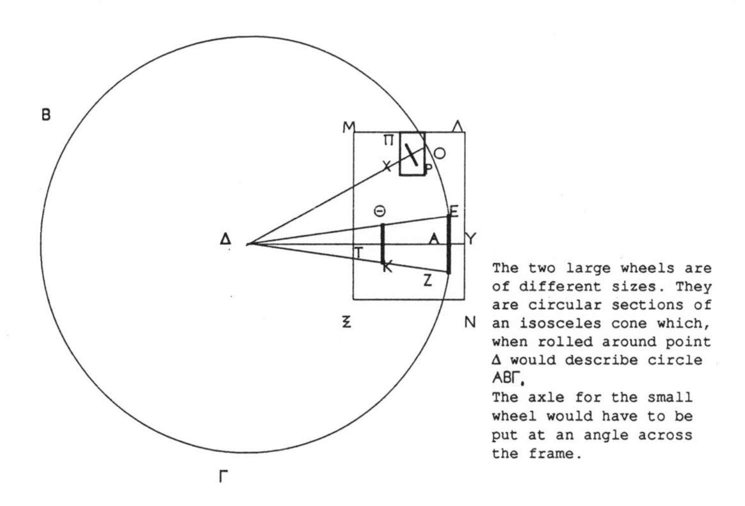

Figure 3 Circular motion.

whose vertex is the point Δ. 2. It is clear that if these isosceles cones are rolled around, they trace circles, while keeping their vertices motionless. For [a cone] laid on a level surface and moving along the line of its side is in balance with itself, because it is bisected by the plane perpendicular to the horizontal which is produced through the side. When it rolls, under the impulse of some external force, each of the semi-circles on its visible surface bears down in the same direction with an equal force upon the remaining semi-circle of the same circle, and thus this also moves. 3. After the semi-circles are considered up to the point of the vertex there is no semi-circle left, nor anything else with dimensions. For this reason, since the potential for movement, by which one part may overpower something set against another part, does not exist, it is impossible to move the vertex in the process of revolving the cone, unless the impetus comes from a forward thrust.[15]

9 1. The motion of the base around a rectangle will be effected as follows: let the base be ΑΒΓΔ, in which let there be an axle ΕΖ with the wheels ΗΘ and ΚΛ attached to it; and let there be a third wheel, ΜΝ, and let the movement and return occur by means of all three wheels, as has been stated. Then let there be a second axle, ΞΟ, with the wheels ΠΡ, ΣΤ and likewise ΥΦ attached to it.[16] 2. But let the axle ΞΟ be above the axle ΕΖ, keeping a sufficient distance from it. And let the wheels ΠΡ and ΣΤ have the capacity to be raised and lowered with the axle ΞΟ – as we shall describe next – and the wheel ΥΦ similarly. Then if we lower the wheels ΠΡ, ΣΤ and ΥΦ so that they rest on the floor, the wheels ΗΘ, ΚΛ and

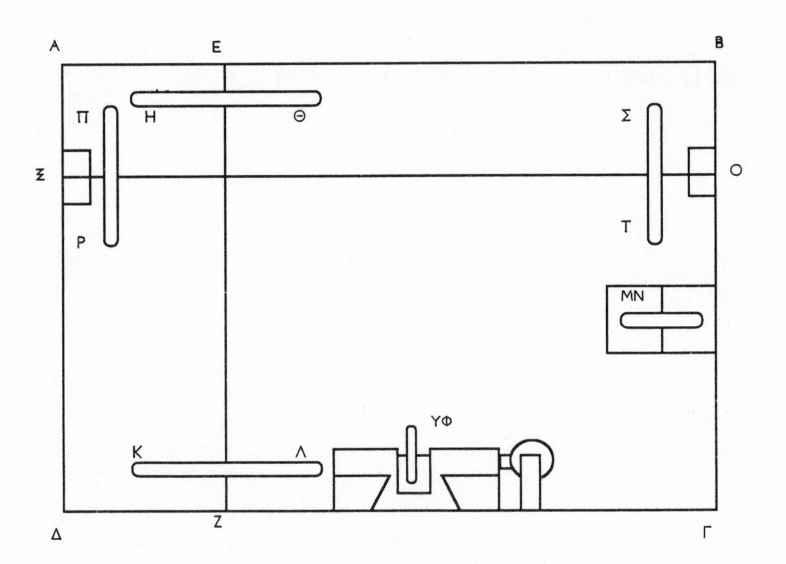

Figure 4 Arrangement of two sets of wheels inside the base for right-angle turns.

MN will be raised from the floor, and the base will move on ΠΡ, ΣΤ and ΥΦ. 3. And when the axle ΞΟ is drawn up so that ΗΘ, ΚΛ and MN rest on the floor again, the base will traverse another side of the rectangle on them. Then, when it stops again, ΠΡ, ΣΤ and ΥΦ will be lowered, and the base will be driven along the next side of the rectangle on them. As these alternations are repeated, the base will travel around the rectangle as often as we want. 4. But it will make individual movements as we may choose by means of the winding up and releasing of the cord. So that the heavy counterweight does not sink too fast in the tube and move the base too rapidly, we shall put something light, small and clinging into the tube, such as millet grains or mustard seed, against which the weight will press down. 5. We shall pierce the bottom of the tube with a suitably sized hole, which will be opened and closed by a lid fastened to a string whose outer end is visible to us through the hole so that we may grasp the string and surreptitiously open the lid whenever we want the base to be moved. As the millet gently dribbles out into the catchbasin below, the base moves. 6. But so that the sudden opening of the stopper will not jerk the base, the string will have a little slack to it, so that as the dribbling of a little millet takes this up, the base will begin to move.

10 1. We shall now describe how the three wheels are alternately raised and lowered. Start with the three aforementioned wheels AB, ΓΔ, EZ. The axle of AB and ΓΔ is ΗΘ. Therefore it is clear that the axle-pins at H and Θ are inserted into sockets in the sides of the base. Let these sockets be set into small bars, and let the bars be fixed vertically into the sides of the base by means of dovetails. 2. Let the small wheel EZ be similarly set in a vertical bar-frame running down through a dovetail to the side of the base facing EZ. Then let this bar-frame be H'Θ', and the slot in it ΛKMN, and let the wheel EZ, with its axle ΞΟ, be in the slot. At the H' end of the bar-frame let a knob, Π, be inserted. In the side of the base facing EZ, let two dowels be inserted as guide-bars – ΡΣ and ΤΥ. Let a screw, ΦΧ, turn on these dowels, and let knob Π project into the spiral of the screw. 3. Then if someone turns the screw ΦΧ, the bar-frame H'Θ' will be raised and lowered by the knob Π.[17] So that this may happen by itself, let a cord be wound around the unused part of the screw with windings and slack parts arranged in kinks proportionate to the distances over which the base is moved. Let these preparations also be made at the two remaining bars, where the axle-pins at H and Θ are. 4. The three screws must be of equal thickness, the loops around them exactly equal, and the slackenings too, so that the three wheels may be raised and lowered at the same time, for thus the motion of the base will be steady and smooth.

11 1. The base can turn, not only in a rectangle, but also in any rectilinear figure: it can even swerve like a snake – and that more easily than by the method described above. 2. Let the base with wheels inside it be ΑΒΓΔ: let two axles, EZ, ΗΘ, be inserted through it, of which ΗΘ should turn easily on its pivots and have the wheel ΚΛ attached to it, while EZ should be joined to the base and trimmed on the lathe to equal thickness.[18] Let two axle-casings, MN and ΞO, be set around this axle so as to turn closely but easily around it, and let both their inner and outer surfaces be worked on the lathe. Let two wheels of the same size, ΠP and ΣT, be attached to these axle-casings. 3. Then if a cord wound around each axle-box is attached to the counterweight in the tube, it will follow that, when the counterweight sinks as the strings are unwound, the wheels will turn

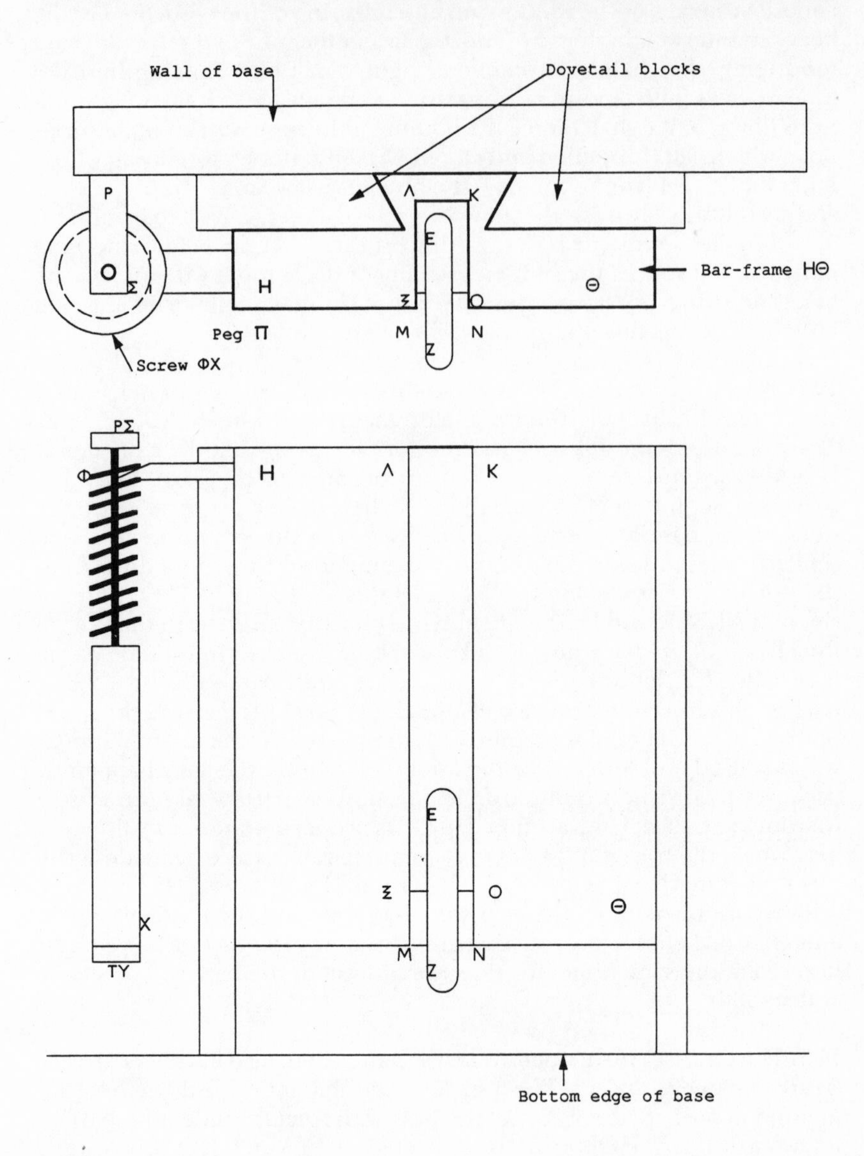

Figure 5

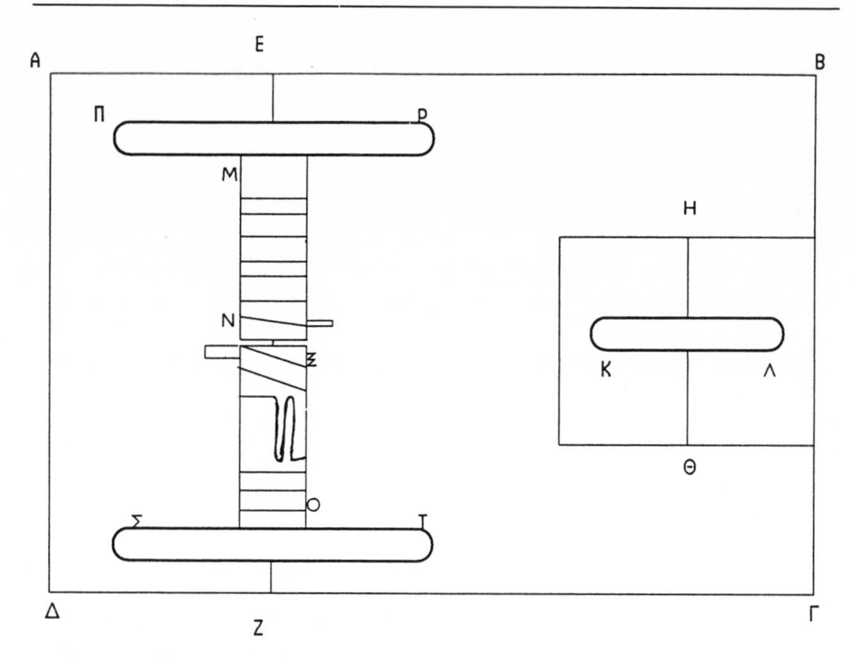

Figure 6 Swerving motion.

with the axle-casings, and thus the base will travel in a straight line, and wheel ΚΛ will turn with it. 4. If, therefore, of wheels ΠΡ and ΣΤ, ΠΡ remains motionless because the cord has some slack there, the wheel ΣΤ will revolve and turn ΚΛ with it, to the point at which the counterweight takes up the slack of the kink on the axle-casing MN. Then, when the cord has tension again, the wheels ΠΡ and ΣΤ will both turn at the same time, and the base will be conveyed along the straight line determined by its own turning.[19] 5. The aforementioned slack, then, must be such that the base turns in the direction we want it to move. Let the same arrangements also be contrived for the wheel ΣΤ. When the cord is wound all the way, with kinks appropriate to those aforementioned straight lines along which the base must be moved, things will happen as described above. 6. We shall have to determine the lengths of the loops and kinks by experiment, because we begin to wind the cord from the spot where the base is to stop; for as we move it by hand in the opposite direction to that in which it will be driven, we shall wind the cords and pay out the slack. For in this way, after the base begins to move, it will stop as it ought at the spot where we began to wind the cords. 7. [It is better that the wheel ΚΛ, also, should be put around the axle ΗΘ on an axle-casing, and that the axle should be similarly fixed to the base just as EZ is, and that the cord wound around wheel ΚΛ's axle-casing, allowing for slack, should be attached to the counterweight; so that whenever we want the base to turn, one of the wheels ΠΡ, ΣΤ will stop, obviously because there is some slack at that spot in the cord round its axle-box; but wheel ΚΛ will roll with the remaining wheel, until the base makes the necessary turn, and then, when the cord with the slack becomes taut again, and all three wheels are moving at the same time, the base travels along its straight path.][20] 8. Then, since the axle-casings which hold the wheels [and are set around the axles] will experience some difficulty in their movement because the whole weight of the base rests upon them, it is a good idea to have all the circular movements of the automata revolve around pivots. We shall do that in this way. 9. Let a right-angled partition be attached in the same manner as the base to the axle holding the two wheels.[21] Let there be sockets on either side of it, in which the pivots may be anchored. Let there be two axles, each with one of the wheels fixed to it, and each set on pivots between the aforementioned right-angled cross-bar and the sides of the base, so that the wheels rest on the floor and each of them rotates around its own pivots. 10. [But let the cords around the axles be double, so that they take the wheel in the middle and turn it evenly. Let the other axle be moved in the same way as these, at the front of the base, so that the base will again ride on three wheels; and let the cord similarly be double around this axle, getting a central grip on the wheel.][22] 11. Then when the cords are again wound crosswise as many times as we want, with the slack distributed as we decide, as stated, the movement of the base will also be as we determine, and smooth and easy on account of the pivots.

12 1. I consider that enough has been said about the motion and return of the base. Now we shall speak of mechanisms beyond that motion. First we have the movement which facilitates the lighting of fire on the altar. 2. It works like this: let there be an altar ΑΒΓΔ, of very thin bronze or iron plate, with a hole, E, in the middle of the hearth. Under this, let there be a small metal plate, ZH, which can slide like the lid of a small box, covering hole E; and let a thin chain, HΘK, be attached to this, round a small axle set inside the altar and revolving freely.[23] 3. On the axle, let there be fixed a cord going back to the counterweight. This cord will be pulled by the counterweight after the base has moved forward, and will turn the axle and slide the plate: all these events will occur in order when the loop has dropped off the peg. Set a fire-grate, MN, under the hole E with its flame directly under the hole; and, as stated, place easily lit kindling on the altar. 4. Woodwork shavings are better than anything else for this. Then, whenever the base stops after being moved forward, the cord stretched from the chain HΘK will draw back the plate ZH so that the aperture is opened and the flame shoots up to light the altar.[24] Make the same arrangements on the other altar as well, except that the slack in the cord ought to be greater than that discussed here, so that during the subsequent movements the cord will be pulled taut and light the second fire.

13 1. After the sacrifice, milk must spurt from the thyrsus and wine from the chalice. 2. This is done as follows: a cylinder is fixed under the Dionysos' feet, with two holes near each other in its upper surface, from which small pipes lead to the inside of the Dionysos, one connected with the thyrsus, the other with the chalice.[25] 3. Let the Dionysos' base be AB, the cylinder connected to it, ΓΔ, and the holes in the cylinder, E and Z. ZH and EΘ are the small pipes extending from the holes – ZH to the thyrsus, EΘ to the chalice; and let the round knob on top of the shrine be KΛM.[26] Inside this, let there be a container, NΞ, with a partition O in the middle. And from container NO let a pipe ΠΡΣΤ connect with another cylinder, ΥΦ, which fits around the cylinder ΓΔ so as to be airtight, and is attached from below to the platform on which the shrine is set up. 4. Then put a hole, T, in line with hole E; and from the container ΞO, another pipe ΧΨΩϛ, likewise bearing towards cylinder ΥΦ. Let a hole ϛ be aligned with Z. Then if someone pours wine into container ON and milk into ΞO, since the holes E and Z are aligned with T and ϛ, the wine will be conducted into the chalice and the milk into the thyrsus. 5. Then, to hold the liquids in at first, let there be a valve, φT', which, as stated, closes off the liquids by means of a stopcock, A', around which must be wound a loop of cord with a some slack to it, attached to the counterweight, so that when it is pulled at the right moment, it will turn the stopcock and the liquids will be released. And when the Dionysos is rotated again and the other altar is lit, the wine and milk must flow again: and thus the stopcock gets turned 180° in two stages.[27] 6. Make two other holes, B' and Γ' in line with holes T and ϛ, and connect a pipe B'Δ' from B' to ΡΣ; and from Γ', run another pipe, Γ'E' to ΨΩ. Then when the Dionysos is rotated holes E and Z will line up with B' and Γ', and the bolt φT' will be opened again and the wine and milk will flow in the same way. [The bolt will be opened when the other string pulls the stopcock to the other side.][28] 7. You must put the pipes ΡΣ and ΨΩ through one hollow column in the shrine from below so that they will be invisible. The Dionysos is turned with the Nike on top of the knob in this way: place an axle ϛ'Z' so as to be attached to the Nike through the knob and freely turning on the axle-pin Z'. Wind a cord around it through the pulley H' and extend the string to the foot of the shrine, through the pulley Θ' to the projecting part of the pipe ΓΔ. 8. So, if one rotates pipe ΓΔ, it will unwind the cord around axle ϛ'Z' and turn the Nike and the Dionysos at the same time. Let them turn in the same direction. Make axle ϛ'Z' be of equal thickness with pipe ΓΔ, so that the Dionysos and the Nike come back full circle without changing their relative position. To make this happen by itself, wind another chain around the projecting part of pipe ΓΔ and through pulley M_A to the weight M_B.[29] 9. Let a ring fastened firmly to the weight be controlled by a catch and a trigger, just as is done on catapults, so that when the trigger is released by a cord, the dropping weight will turn the Dionysos and the Nike. And let the cord H'Θ' be hidden by another column, as explained for the pipes.[30]

14 1. After the Dionysos first pours the libation there should be a banging of drums and cymbals. This is done as follows: in the lower part of the pedestal, where the wheels are, is placed a box containing lead balls, which roll out, one by one, on to the floor of the pedestal. On the floor is a hole the right size for the balls to fall through easily, and this has a closure which is opened as needed by a cord.[31] A little drum is placed at an angle under the hole, and a little cymbal should be fastened to it. 2. Then, as they fall, the balls will strike the drum first and bounce off it on

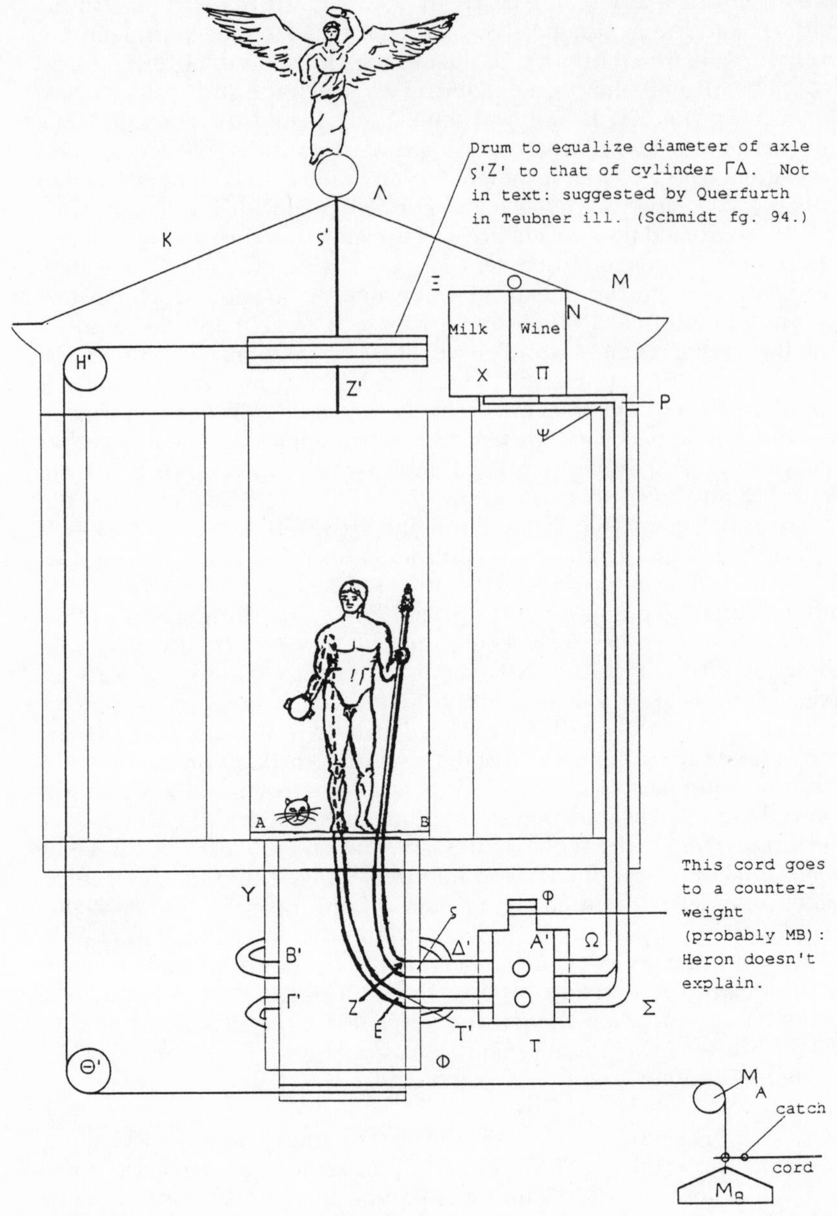

Figure 7

to the cymbal to complete the sound effect. A partition can divide the container into two chambers so that there are balls in each one and those in one section produce the first set of sounds and those in the other, the second, when the stopper there is opened in the same way.[32]

15 1. Next you must put garlands around the colonnade on the base. 2. This is accomplished as follows: devise a frame, ΑΒΓΔ, set on top of the four columns, containing another frame, ΕΖΗΘ, so that the space between the two frames is empty on the underside.[33] A wreath made of four garlands, braided however you like in order to please the eye, is folded and hidden in the aforementioned space between the frames,

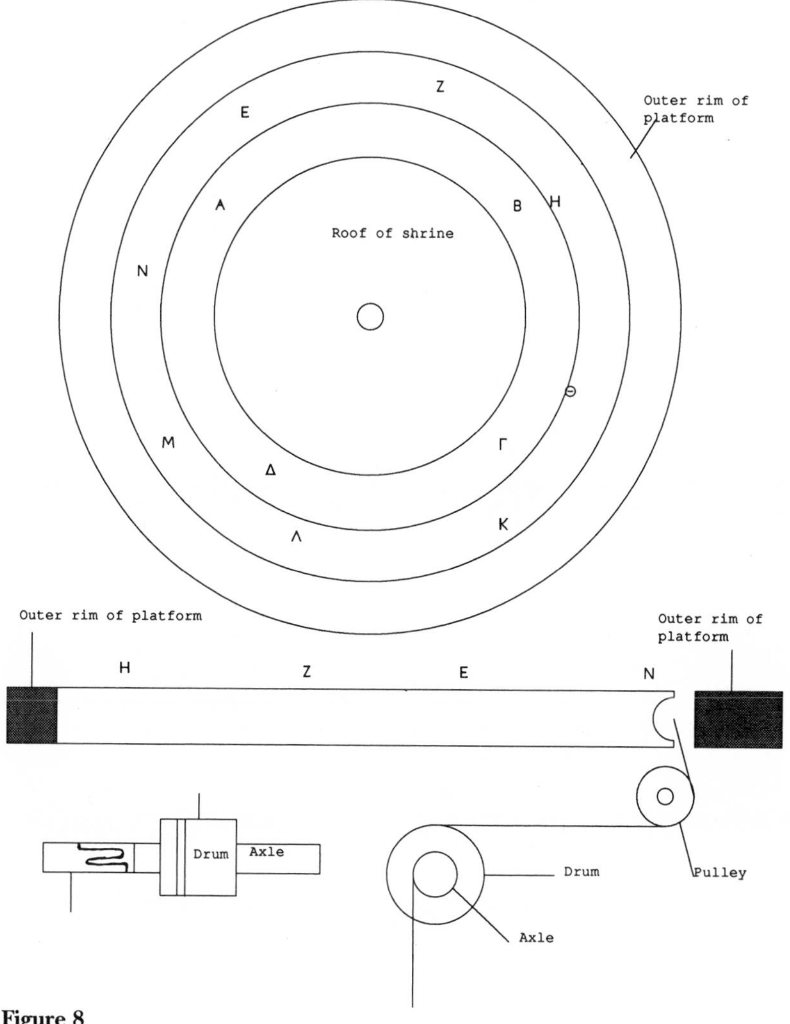

Figure 8

with its upper ends fastened to the frame. 3. And to keep it from falling down by itself, a long board must be fitted to the space between the frames along each side of the frame, so as to cover the wreath like a lid and secure it to the upper part. But so that the boards do not swing loose by themselves, they should have easily opening hinges on one side, toward the inside of the frame, so that when the boards are closed they are held up on the other side by a reversible hook, to prevent them from opening.[34] 4. Then a loop of cord is wound around the hook from the other side, so as to fall off when the cord is pulled tight and the hook is turned. And thus the wreath is lowered. The wreath will have little lead weights on its lower side to make it drop down quickly.

16 1. It now remains to show how the Maenads dance at the appropriate moment. This effect is produced as follows: let the round shrine containing the Dionysos have a round foundation which is smooth along its upper surface. Let this, then, be ΑΒΓΔ; and around it set a felloe ΕΖΗΘΚΛΜΝ, fitting snugly to the foundation, so as to be rotated around it easily. 2. Using a lathe, cut a groove around the circumference of the side ΚΛΜΝ; wind a cord snugly [into the depth of the groove][35] and let one end of the cord be pushed into the depth of the groove by a wooden peg, to stop it from being pulled out any further. 3. Pass the other end around a pulley to the lower part of the frame, and wind it into another groove, on a drum: fit an axle closely to the drum with enough room to turn freely.[36] Another cord should be wound around the axle and attached to the counterweight. It will follow that when the cord around the axle is pulled, the cord from the felloe is wound on to the drum attached to the axle, thus causing the Maenads to dance. Since they must dance twice, the cord around the axle has a slack loop in reserve, so that the Maenads will come to a halt because of the slack; but when it is taken up they will dance again; for the Maenads will be placed on the aforesaid felloe.[37]

17 1. But any cords reaching from the lower pedestal to the counterweight must be hidden from view.[38] This is done as follows: let the mouth of the tube containing the counterweight be ΑΒΓΔ, and insert a partition into the mouth of the tube along the straight line ΕΖ, separating off the narrow space ΔΕ. 2. Then the millet will be put into the section ΕΒ, but the cords from the lower parts will come up through section ΓΔΕΖ and be attached to the counterweight in section ΑΒΖΕ by way of a pulley: thus all the cords that come up from below will be hidden. Then since there are many movements and the wheel-base moves a long distance, the tube must be tall enough, and you must make it so. 3. Now, regarding the distance of the movement, either the two wheels around the axle can be made bigger to increase it or the diameter of the axle can be lessened. For when the axle rotates once the base will be moved as great a distance as the circumference of one wheel. Therefore logic dictates that the wheels be made bigger.[39]

18 1. But it can also be done this way: let the thickness of the axle be AB, and the circumference of the wheel fixed on it, ΓΔ; and set above it another axle, of thickness EZ, turning freely on pins. Snugly fit a drum HΘ to the axle, and let a cord wound around the axle AB be wound around the drum HΘ. Attach the other cord fastened to the axle EZ to the counterweight Λ in the tube, by way of a pulley, K. 2. Then it will follow that when axle EZ is rotated once, a small part of the tube, proportional to the circumference of axle EZ, will be emptied; but the drum HΘ, which is bigger than axle AB, will wind the cord from axle AB once, and consequently axle AB, with wheel ΓΔ, will rotate more than once, and thus the motion will cover a longer distance. 3. However, it must be recognized that a bigger counterweight is needed, since bigger wheels are being moved by smaller ones: these things are done by levers. 4. It is also possible for motions other than the movement of the base to be completed on a large scale by using small radii; for if the cord moving the apparatus of the Dionysos is wound around greater circles, it must then go to the counterweight around smaller axles attached to larger ones, as I explained for the main motion.[40]

19 1. Now, the entrances and exits of the automata, as well as other movements, can be effected in another way. Let the mouth of the tube ABΓΔ be divided by two partitions through the whole height of the tube along the lines EZ and HΘ so that the cords may be drawn up through the space between the partitions and attached to the counterweight. 2. Then the counterweight in the tube ABEZ will control entrances and exits, while that in HΘΓΔ controls the other movements. Let the hole in the bottom of the tube ABEZ through which the millet funnels out be K, and let the hole in HΘΓΔ be Λ. Let each hole have a lid which slides smoothly open and shut. 3. When the wheel-base is about to move, we slide open the lid of hole K; but so that the base will not move too fast the instant it is pulled, there will be some slack in the cord which goes from the wheels to the counterweight.[41] And clearly, after we step back and before the base begins to move, there will be an interval proportional to the amount of the slack. 4. Then, when the base must stop and the other motions must be completed, a cord will be pulled while it is still travelling, and the lid at Λ will be opened. Again, so that no other movement occurs during this motion, the cord extended from the other counterweight (which when pulled will also open the lid under hole K) will also have some slack. 5. And thus the base will stop, and the other motions will come to completion. Then when the base must be moved back again, the other cord will pull the lid at K, and it will be opened; and thus the base will make its exit.[42]

Book II: On Stationary Automata

20 1. I believe I have sufficiently discussed in the foregoing whatever needed to be said about moving automata; for I have recorded schemata which are viable, safe and novel compared to those written down by my predecessors, as is obvious to anyone who has tried the earlier plans. I would also like to write something original about stationary automata; and I find, of my predecessors' writings, none better and more suitable for didactic purposes than those of Philon of Byzantium. 2. The plot and representation portray the legend of Nauplios, and contain many varied scenes, which are not clumsily managed except for Athena's crane; here, for some reason, he made the mechanism unnecessarily difficult, because it is quite possible for her to appear on stage, and then disappear again, without a crane. For the figure of Athena can be moved on a hinge at its feet, and be laid horizontal at first, so as not to be seen, but then as though drawn up by means of a cord, it can appear, upright, and then be hidden again as though drawn down by another cord.[43] 3. Further, although he promised, in addition to this, to make a lightning bolt fall on the figure of Ajax, and to produce the sound effect of thunder, he did not explain how; for in reading many schemes we do not find this discussed. Perhaps some will think that in criticizing Philon I am accusing him of being unable to finish his project, but this is not the case. 4. There are so many devices promised in his scenario that perhaps he forgot to write this one out in full; for a container full of lead balls with a hole in the bottom could be opened as needed, letting the balls fall upon a tightly spread piece of dry leather to produce the sound of thunder. For even in full-sized theatres, when such a sound is needed, containers of weights are released, so as to strike dry hides, as I said, and because the hides are stretched tight they make a drumming sound. 5. I am quite happy about all the other things that happen severally in the Nauplios scenario, as explained in order and methodically by Philon. Therefore at this point I am not rejecting his writings about the things I am discussing, since I believe that readers receive the greatest benefit in this way, when things said well by older authors are presented to them accompanied by comparisons or corrections.

21 1. Let us begin by discussing the construction of toy theatres. Now, compared to the construction of the moving automata, the making of [stationary automata] is very much more reliable and risk-free, and can be more convincingly demonstrated.[44] The problem is this: to make a toy theatre which is set upon a wooden pillar open by itself, so that the figures painted in it appear to be moving in a manner consistent with the demands of the scene being presented, and when the doors close again automatically, and open again after a very short period, other painted figures appear in it, and so far as possible these, or at least some of them, move; and this happens several times. 2. And off-stage either a crane is seen being raised and turned round, or some other movements. Now, this is the proposition: but the most elegant handicraftsman is the one who invents the most polished scenario. Therefore I am going to propose one such scenario, the one I consider best, and then explain how it works. The example of one toy theatre will suffice: for these things are all managed by the same methods, just as I demonstrated for the moving automata.[45]

22 1. The ancients used a simple device: when the theatre was opened, a painted mask appeared within it. This moved to the extent of opening and closing its eyes repeatedly. When the stage, after being closed, was opened again, the mask was no longer to be seen, but painted figures set in order for some story. 2. And when it was closed and opened again, another array of figures would appear, illustrating each story in turn. The result was that only three movements occurred in the theatre, one that of the doors, another of the eyes and the third of the backdrops. Our contemporaries, however, have set sophisticated stories on these toy stages, and have made use of varied and unconventional movements. 3. As I proposed, I shall speak about one presentation which seems superior to me. The story adapted in it was the one about Nauplios. Scene by scene, it went like this. When the stage was first opened, twelve painted figures appeared, arranged in three rows. They were made to portray some of the Greeks repairing their ships and busying themselves about launching them. 4. These figures moved, some sawing, some working with axes, some with hammers, others using bow-drills and augers, making a lot of noise, just as would happen in real life. After sufficient time had elapsed, the doors were closed and opened again, and there would be another scene: the ships would appear, being launched by the Achaians. When the doors were closed and opened again, nothing appeared in the theatre except painted sea and sky. 5. After a short time, the ships sailed out in line, some hidden, some visible. Often dolphins swam alongside them, sometimes diving into the sea, sometimes visible, just as in real life. The sea gradually turned stormy, but the ships ran on. However, when the stage was closed and re-opened, none of the voyagers was visible, but only Nauplios holding a torch and Athena standing next to him; and a fire was lit above the stage, as though it were the flame of the torch. 6. And when the theatre was closed and opened again, the wreck of the ships appeared, and Ajax swimming; [and Athena][46] was lifted on the crane above the stage, and with a peal of thunder a lightning bolt fell in the theatre itself, upon the figure of Ajax, which disappeared. And thus, as the theatre closed, the story reached its climax. Such was this presentation.[47]

23 1. As I wrote, you must build your toy theatre as big as you want it, constructing a suitably sized box out of very light boards. The boards should be one-sixth as wide as the length of the longer sides. 2. You must fit the floor of the stage to the middle of the box, and under the lower part of the box attach a hollow compartment, invisible from the back, to which the doors are attached, with pivots under it of such length that when they are turned from below the doors open and close. 3. Then let the box, which is visible from the front side, be AB, and the pivots going down from the doors be Γ and Δ. So if someone turns the pivots on each side by hand, the doors will open and close. Then, to make this happen automatically by means of a cord, when this cord is pulled by the counterweight which rests on the sand in the tube, I set a perfectly round, rotating axle, EZ, athwart the pivots and a little bit behind them. 4. I

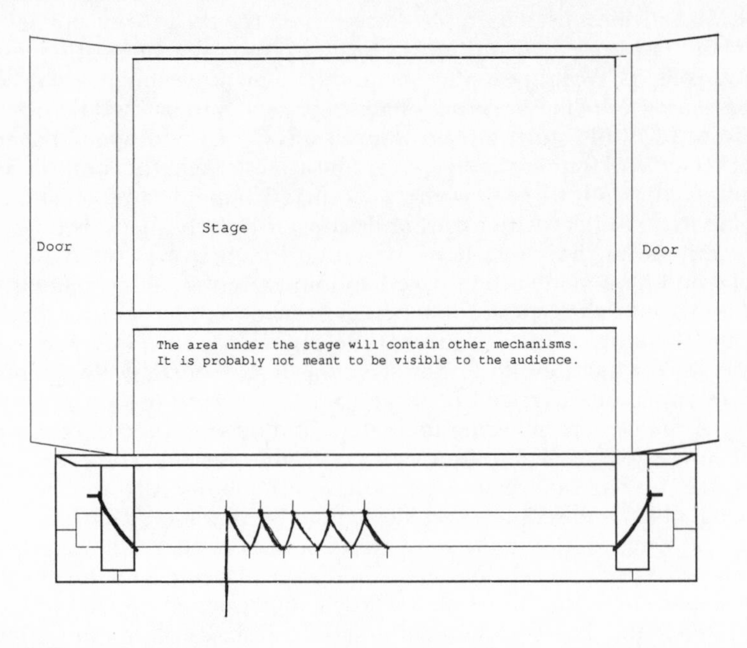

Figure 9

drilled each of the pivots and, taking a cord, wound it in two parts; I inserted one strand into the hole, drove a peg in and glued it, thus locking the cord in, so that it would not be pulled out, but remain fixed in place. After doing this, I put the other ends around the axle, one going over it at ΓΔ, one going under it at EZ. 5. After drilling the axle at each end in the same way, I locked the cords in quite securely with pins, one at E and one at Z. The cords will turn the pivots and open the doors. When I turn the axle the other way again the cords are pulled back up and so the doors will close. 6. In this way, then, both doors will be alternately closed and opened with one motion. Then, so that this would be done automatically by the counterweight, I fixed knobs to the axle – those going up, H, those going down, Θ, and I took the cord, and after measuring its length against the tube containing the sand and the counterweight, I made loops at appropriate intervals. 7. Let the cord be K, and the loops be Λ. Then I put the first loop from K around the first knob from E, at point H, and the next loop around the lower knob Θ, and thus glued all of them in order around the axle EZ with wax and resin.[48] This gluing is masked.[49] 8. I also glued the slack part of these loops to the axle, so that none of them would get in the way and throw off the mechanism. Thus when the end of the cord, on which K is, is hung from the counterweight and pulled gently, it will open and close the theatre, establishing the timings and intervals.

24 1. These events, then, happen in this manner. When the first scene opens, we must <show>, how ... figures of men doing carpentry work appear on the stage; and we must demonstrate how they receive their motion.[50] You must paint every other part of the figures on the back wall of the stage in the most plausible representation; however, the right arms should not be painted on the theatre, but made out of thin, lightweight pieces of horn, whittled down very finely and attached so that they fit flush and do not leave an obvious gap. 2. And the ship's gear on which they are working must also be of horn and set in their hands, and the hands must be painted the same colour as the rest of the body, and the ship's gear as is appropriate. So, let there be an arm, AB. I drilled this at the shoulder and made the hole square, as has been illustrated; then I took a horn peg, squared it off, fitted it to the shoulder and glued it on, making the rest of the peg round and very smooth. 3. I drilled a hole through the right shoulder, and then drove a peg in hard until the little

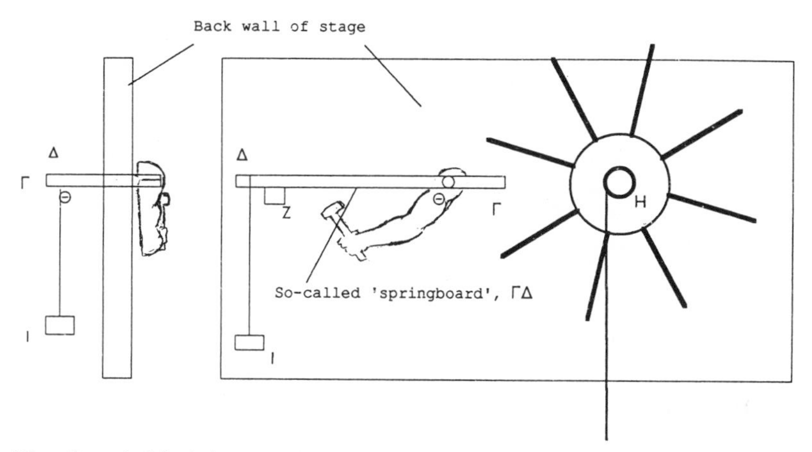

View from behind the stage.

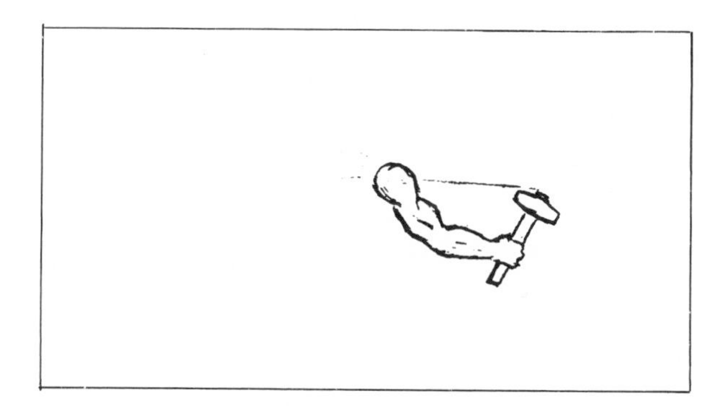

View from front of stage.
Figure 10

arm was set snugly against the figure. Thus if we take the projecting part of the peg in our fingers from the back of the theatre and turn it, the hand will move. To let it be moved automatically by the counterweight, I make a bar ΓΔ, drill a hole in it at Θ and attach the peg projecting from the arm to the back of the theatre fixing it snugly into the hole in the bar and gluing it in, so that when the bar is moved, the peg attached to the shoulder will also move. 4. This bar is called the springboard.[51] After piercing one end of the springboard, I attached a cord and hung a lead counterweight, I, from it, and fixed a peg, Z, under the very end of the springboard, so that the end of the springboard would come to rest on it. So, if we push the Γ end of the springboard down with our finger, the Δ end is raised with the counterweight, and if we let go, it will fall upon the peg as the counterweight pulls it down, and make a noise. 5. It will also impart motion to the arm, on stage. Then, so that it will move regularly and automatically, I set a spiked wheel turning round a pin fixed firmly in the back wall of the theatre. The wheel will have fixed to it a pulley, H, around which a cord is wound many times and then fastened to the counterweight, so that when the counterweight is applied a little bit, it turns the spiked wheel and the wheel, in its rotation, strikes the springboard regularly. 6. The end part of the cord is wound in loops around the knob where H is. When the hand is no longer moving, you will recognize that the cord has fallen off the knob.[52]

25 1. This is how the scene with the woodworkers is handled. But when the scene changes after these events, the woodworkers must be replaced by the ships being launched. 2. I will now describe how this is done. You must take a thin and finely woven piece of linen cloth the same size as the back wall of the theatre, painted with thin white pigment, so that it can be rolled up easily; and you must paint the ships being launched on it; and putting it into the theatre, nail the upper part of it to the back wall of the theatre with small tacks, under the actual side of the box; and fasten a bronze rod of consistent thickness to the lower edge of the cloth, so that by winding 3. the cloth around the rod up to the top of the theatre and gathering it up we may contain it under the side of the box;[53] then when we wish, we can let the cloth be unrolled from there by the weight of the rod, so that by descending thus suddenly, it covers the painted figures on stage. 4. This, then, must happen automatically when the stage is closed. But at first it must remain tightly rolled up above, which is done as follows. When it is rolled up securely towards the top and placed under the side of the box, a hole is drilled below the covering near it, in the back wall of the theatre, and a loop of cord is pushed through the hole from the back part of the theatre to the front part, until it projects far enough and is attached securely to a peg. 5. The exact distance will be determined by specific needs. Opposite this hole, I drill a hole, wider than the one behind, in the back wall of the stage, next to the side of the box, and file it all the way through, to make it smooth. The roll of cloth must be wound tightly next to the back wall, wedged into the loop; and a little pin must be pushed through the little hole in the side and driven upward through the loop.[54] 6. Then the roll of cloth remains bound up by this loop. But when the objects on stage must be masked, after the doors are closed, pull the cord which is attached to the pin and goes back to the counterweight. Make all the backdrops this way, roll them up and set them in order close by each other, to unroll from the top, each of them with its own loop and pin. 7. The space the rolls of cloth occupy should be blocked by boards, so that they will not be seen. The short board becomes the door lintel. It must be made into an architrave with carved mouldings, so as to have a finished appearance.

26 1. This is how these effects are produced. When the next scene opens, nothing should be visible except painted sky and sea, and then the ships sail by. So we shall also engineer the appearance of the voyage as follows: the stage will have concealed hollow spaces on each side of the doors by the hinges, finished on the front to look like pilasters. 2. In these spaces are placed concealed boards with square rods of equal dimensions inside them, standing upright, with their angles bound. These will be of fir, so that they will not warp because they are too thin. And set on the underside of these will be rounded bronze knobs, with sockets set beneath them so that they can turn easily; while at their top ends, the rods are worked round and smooth. 3. And at the top they are inserted through the pierced side of the base, so that they neither bind nor turn too easily. After you take care of this, you must take a very thin piece of what is called royal papyrus and cut a length of it equal to the height of the back of the stage, reaching up as far as the rolled-up cloths; then you must cut off the roller of the papyrus and glue it to the rod on the right side of the stage, 4. so that the rod is glued centrally to the roller and thus, when the part [of the rod] protruding above the stage is turned, it winds the papyrus around the rod after the scene is over. [This is turned until the whole back of the stage is covered by the papyrus, which will (not) happen until you bring it up to the other rod; and only the back wall of the stage, up to the rolled-up cloths, must be covered by the papyrus when you bring the latter to the rod, and so you fill up the space, and if there is some extra, cut it off.][55] 5. Next you must glue a very thin worked rod under the edge of the papyrus. Let the cloth rolled up near the pilaster be hidden so as not to be seen when the stage is opened. So I fasten thin cords to the little rod glued near the edge of the papyrus – one below near the pilaster of the stage door, and another above, near the lintel – and extend it to the other rod on the left side. 6. Then if we rotate the rod it will pull the cords: for the cords, being attached to the edge of the papyrus, will be wound up, and the papyrus will follow them. Then, when the stage is closed, turn the rod far enough to make the papyrus, which will have sea and sky painted on it, mask the whole back cloth. You must see to it ahead of time that the papyrus moves into place automatically, and, when the heavy counterweight pulls it, is deployed quickly to produce the effect of a great number of ships sailing by. 7. Let the theatre as seen from behind be ΑΒΓΔ, and the upper end of the rod around which the papyrus is rolled <have>[56] a rounded cylinder ΖΗ, and against the stage, above the springboards and spiked wheels which move the hands, at a slight distance, I set a small drum, ΘK. Let the drum have a sheave slot lathed around its rim. 8. Around the axle of the drum, I set another small axle, M, snugly attached to this axle, so that it is turned at the same time as the larger drum. After winding enough cord to unroll

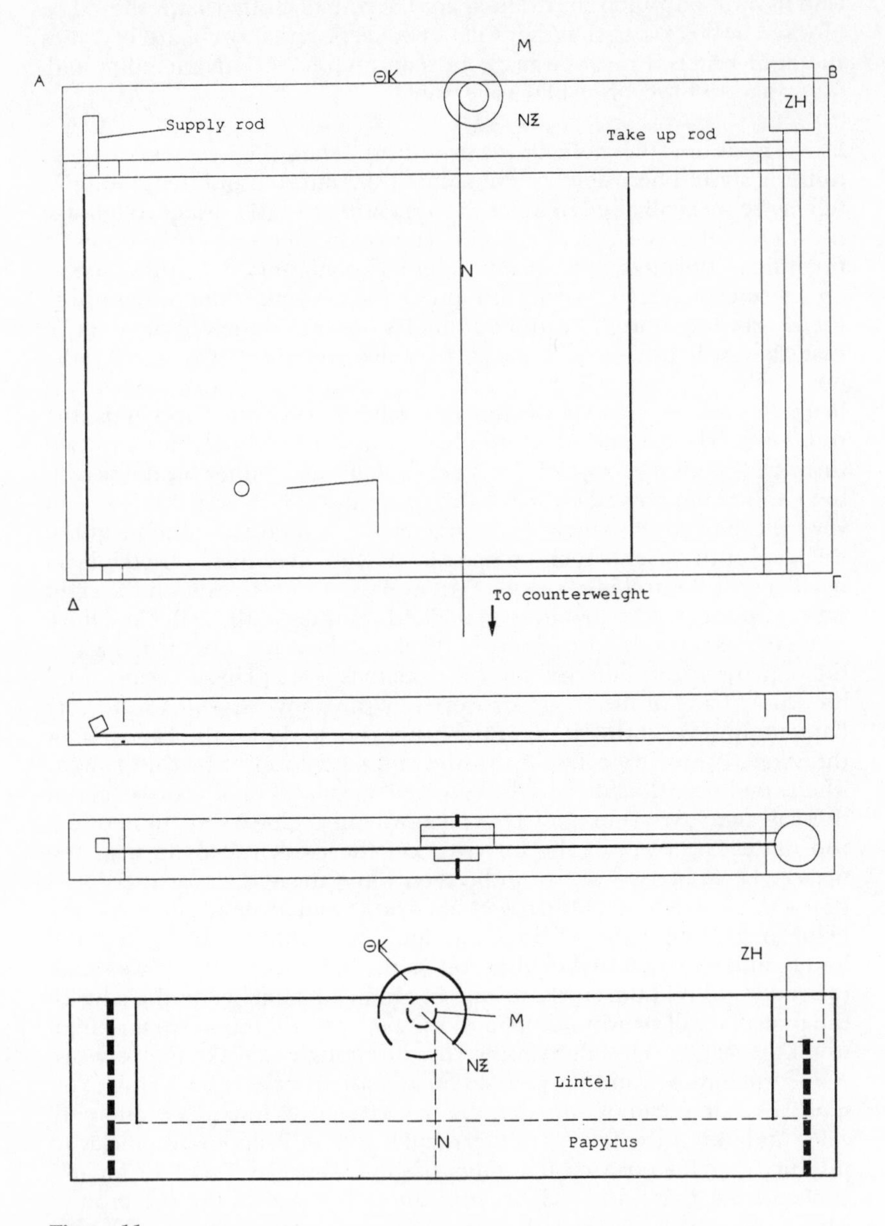

Figure 11

the papyrus around the cylinder HZ, I attach the cord back around the other pulley <on> the first drum [ΘK]. I wind around the axle M, <which turns with the first drum> the cord which goes back to the counterweight: let this cord be N. 9. It is clear, then, that if the cord is pulled only a short distance by the counterweight, a great part of the papyrus will be wound up, and quickly. The axle on which the drum revolves is NΞ. The spiked wheels and the drum must move without encumbrance.

27 1. So this is how the effect of the ships sailing past is created. Dolphins can also be made to dive and surface alternately, in the way described below. In the lower side of the main box, which is flush against the smaller compartment, a little way from the hinges, I cut narrow slots, like tenons, which allow light through to the lower compartment. 2. I took a board and drew little dolphins of the size I wanted, cut around their outlines, and filed the edges smooth. Now, let a little socket be attached under each dolphin's chest, in which I inserted an iron pin thus <anchoring> this too to the dolphin's chest.[57] Let there be a pulley firmly fixed into the slot on one side, as described below. The slot in the surface is AB, the axle is ΓΔ and the pulley is EZ. 3. I pierce the axle, then, at Θ, facing the slot, and fix the dolphin's pin into it. So if one spins the wheel by hand the dolphins alternately dive through the cutting into the lower compartment and leap up on stage. 4. So to make this happen automatically, I arranged loops in a cord and wound it around the knob Z, on the pulley; and after winding up the pulley, I extended the cord back to the counterweight. And thus the little dolphin will be fixed to the axle, so that <the pulley> where K is, is at right angles to the axle, and the axle, ΓΔ, is at right angles to the lower chamber.

28 1. So when the sailing scene is over the doors will close again, and the cord, stretched taut, will pull out the pin and unroll the backdrop on which Nauplios is depicted holding up the torch, along with Athena.[58] And when the stage is opened again, the ships will not be visible, but instead those things mentioned above. It will be necessary to light the torch immediately. 2. Therefore we shall make the mechanism for the torch as follows: we will have a board on top of the architrave and triglyphs, which casts a shadow over the whole stage, and masks the cylinder that activates the sailing scene, the device which lights the torch and the lifting of the crane, so that none of them will be visible from the front part of the theatre. 3. But so that the board is not left sitting there inexplicably, a pediment is added to it, just as to a shrine, and let the projecting sides of the board be coloured black or sky blue. The crane is set next to the cylinder; but the construction of the torch, on the opposite side from the crane, must be carried out as follows. You must make a sort of small box out of thin bronze plate, not with a lid, but open. 4. And this you must stand upright behind the masking board, and nail to the surface of the base. And let the box have its base against the board, with its open side looking out from the board. And from the upper side of the box cut an opening which admits light like a window, so that when the lamp is lit and put in the box, the top of its flame will shoot through the opening to the part above. This done, set the lighted lamp in place

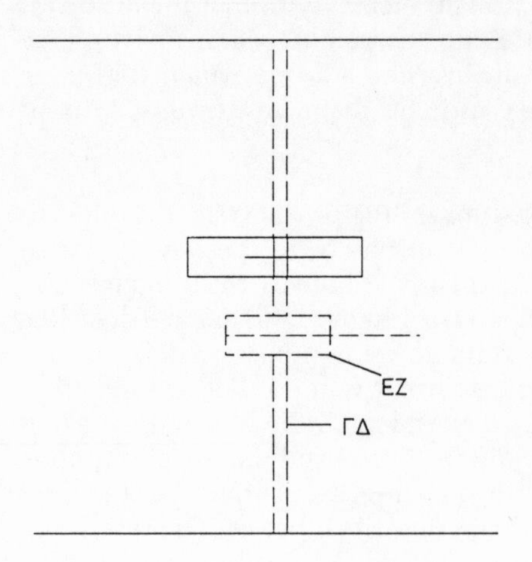

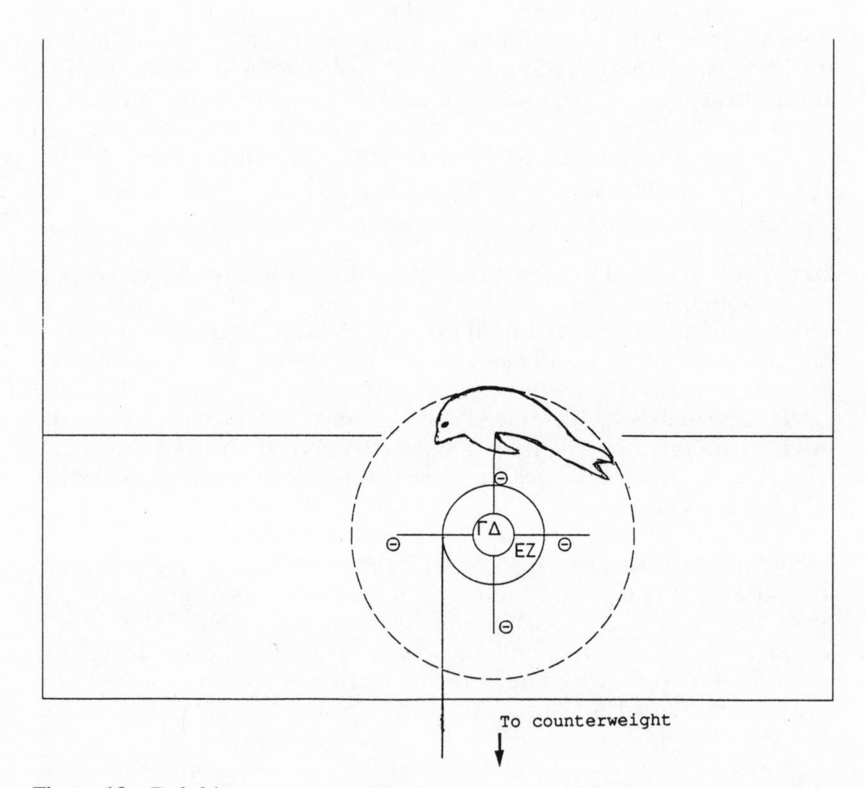

Figure 12 Dolphins.

underneath.[59] 5. We cover the aperture with another bronze lid, thin and triangular, designed to close off the flame. And above the box and its cover I place some extremely dry wood shavings. So when I pull away the small lid which covers the opening, the flame of the lamp will kindle the shavings and they will blaze up immediately. Until the shavings catch fire, the lamp's flame is not seen, being hidden in the box. 6. Besides, the box will have a wooden plug, if we want to make the flame invisible from all directions by covering it up completely. In order for the lamp to remain

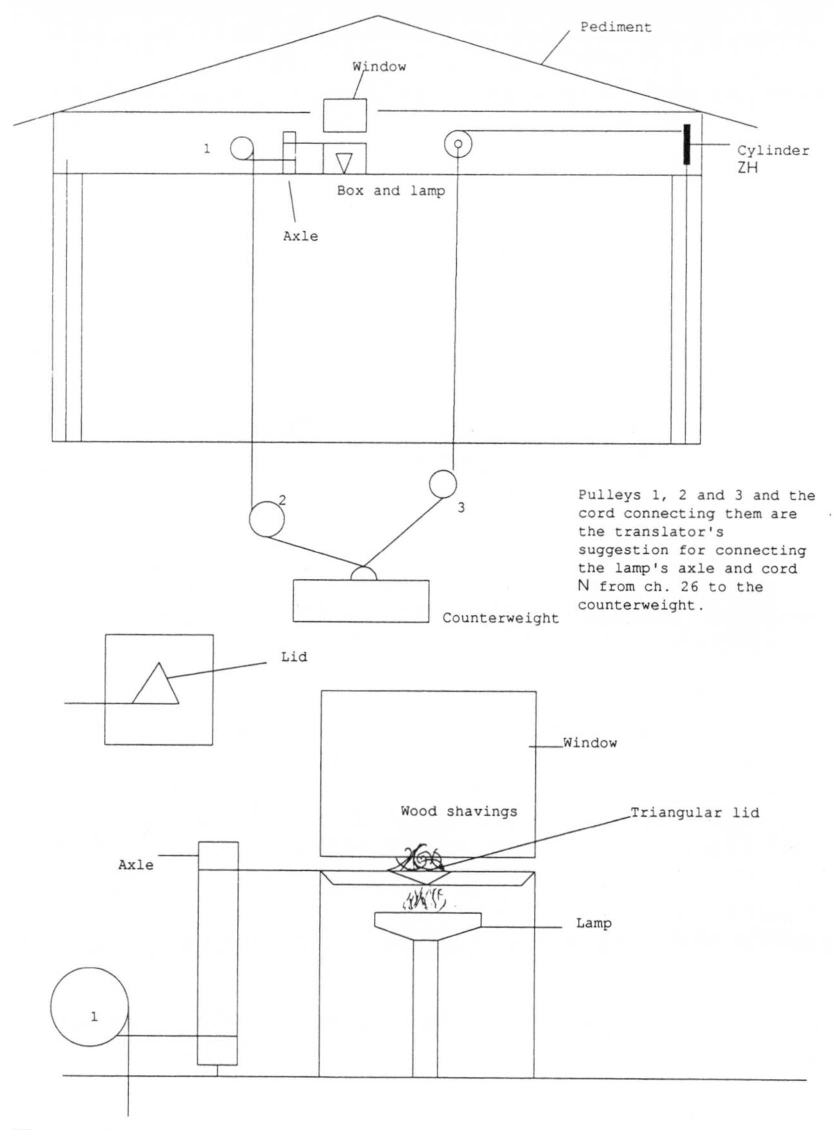

Figure 13

safe and steady in the chamber, let there be a peg sticking up from the bottom, and let the lamp be of the sort which is set on a stand and anchored around such a pin. Then, so that the little lid is opened at just the right moment, I set an axle back a bit from the flame. 7. Then I fixed a little chain to the lid and attached it from there to the axle, so that when the axle is rotated, the chain is wound up and pulls back the lid.[60]

29 1. When the aforementioned things have been seen and the torch has been lit, the stage will be closed again. And when the cord pulls the pin out it will release the cloth on which the wreck of the ships and the swimming figure of Ajax are depicted. And Athena will appear on stage; her base will have pegs at appropriate points. 2. One cord will raise her by pulling from behind her hips, and will keep her balanced. After this cord is released, another, set around her waist, pulls her in a circle, until she returns to the spot from which she started; but when this is let go, another cord will be pulled from in front of her hip and in this way Athena will be drawn out of sight.[61]

30 1. It remains for us to discuss how the lightning bolt will fall on stage and the figure of Ajax will disappear. These effects are achieved as we shall now explain, in turn. The painted figure of Ajax will be set at the back of the stage. In line with this, let there be one slot in the ceiling of the toy theatre and another in the floor, just as we explained for the dolphins. 2. Two of the thinnest gut strings, the kind used in lutes, are stretched from the slot in the ceiling until they reach down into the lower chamber through the cutting at the top of it. To ensure that they are stretched taut in the shrine itself, they are fastened on to two screw pegs in the upper part, so that when the pegs are tightened the strings will have tension. 3. Next, set up a thin and longish board so that it fits easily through the slots and stands behind the lintel but does not protrude past the lintel into the theatre. Next, drill two holes at each end and attach the gut strings to these by means of pegs. Finally a little lead plate is also glued behind the board to counterbalance its weight. 4. Then if we push the board up through the slot by hand, it will fall straight down through the stage, since it rests on the strings. The strings, then, are dyed black, so as not to be obvious. The underside of the board is gilded and made as smooth as possible, and something fiery looking, to represent the lightning, is painted on the upper surface.[62] 5. And this, when it is let go, is projected straight at the middle of the figure, since the strings are taut. But the figure still remains held from above by the pin, along with the cloth, until, at the right moment, the cord pulls the pin and the lightning bolt falls.[63] When the lightning bolt falls, the figure disappears as follows: there is another cloth prepared just like the others which are concealed, but small, just big enough to mask the figure; and the sea and waves are painted on it in the same pattern as they are around the figure. 6. And if any other physical object is visible, it will be added (to the cloth) so that when the figure is covered the scenery will be the same. The cloth must also be coated with a sea-like colour on the back, and to prevent its being seen at all when it is concealed, it is rolled up, and held in place by the same pin that controls the lightning bolt, so that when this is pulled out, the lightning bolt is cast down on the figure, and the figure is covered by the cloth at the same time, so that it looks as though the figure disappears because it is struck by lightning. 7. So this is how these effects are taken care of in the toy theatre. The individual movements of the figures and those of stage action all come about through the same devices, and all toy theatres are similarly operated by these means, except for whatever is altered <because of the story>.[64]

Notes and References

1. The rails fit around the wheels, instead of the wheels fitting over the rails as on a modern railroad.

2. **2** 1–2. Heron introduces the setting necessary to make the automaton work – a smooth, hard, level surface, or a set of rails for the wheels. That he mentions the level surface shows that Heron was concerned about the effects of friction on the wheels; his concern for oiling the axles and bearings (**2** 4) also demonstrates an awareness of friction. So, perhaps, does his use of cords to turn axles (**2** 7); but most of the cords are secured to pegs on the axles

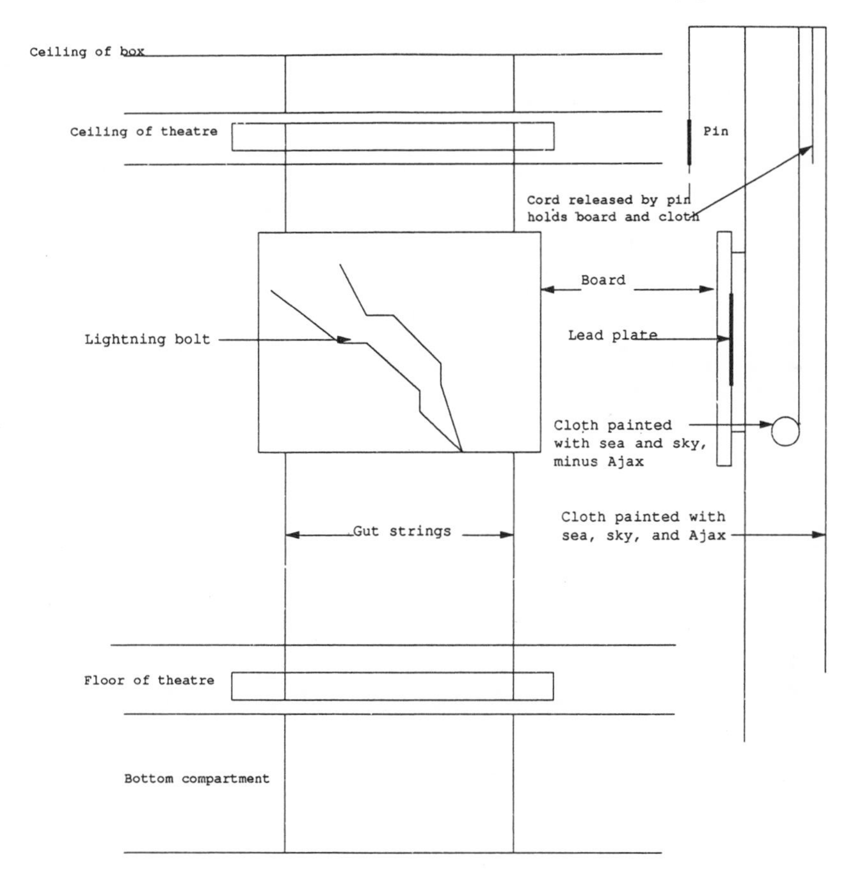

Figure 14

(**5** 4), so the idea may have been more that the cord was pulling on an object to which it was directly attached (the peg), with the shape of the axle imparting a circular component to the motion, than that the friction produced by the cord on the axle was the source of motion. **2** 2–3. The materials used for the automaton are light wood for most of the construction; bronze or iron for axles and pivots and bearings. Heron recommends lead for the counterweight (**2** 6.).

3. **2** 4–5. The oil would have been olive oil. The cord (probably of flax) is pre-stretched and varnished to prevent its stretching or contracting once installed in the automaton.

4. **2** 6. Sinew rope wouldn't work here, since constant tension, not springiness, is what is wanted. ὕσπληγξ (hysplenx), 'springboard', normally means either a twisted strand (i.e. of sinew, used in catapults), or, as here, a piece of wood whose movement is controlled by twisted sinew cord (but see **24** 4.). Ancient torsion-powered catapults were powered by two twisted skeins of cord made from animal sinew, into which the two arms were inserted. When the slider which held the missile (a stone or a bolt) was drawn back, the arms were pulled back as well, further twisting the sinew cord. The rapid untwisting of the sinew would propel the missile. Heron's *Cheiroballistra* describes a torsion catapult (arrow-shooter). The type of springboard referred to here would work the same way as a catapult arm; however, the only hysplenx in this text appears at **24** 4, and is not powered by sinew cord. For ancient catapults, see Marsden, 1969 and 1971, the latter for a text and translation of Heron's *Cheiroballistra.*

5. At **3** 1., Heron suggests dimensions for the base, but he never suggests an ideal weight for the counterweight. This would depend on the force the weight needed to overcome, which is determined in part by the diameters of wheels and axles; cf. Heron's remark at **18** 3.

6. The descent of the counterweight pulls all the cords at the same rate, but the different sizes of the axles – like differential gears – allow events in the moving automaton display to occur in sequence. Heron may have had knowledge of working gears – the Antikythera mechanism, a clockwork calendar dated 150 years before Heron, had a complex system of differential gears for keeping track of astronomical phenomena. This calendar may have been operated by a hand crank: it was not automatic. (See Derek J. De Solla Price, *Gears from the Greeks: The Antikythera Mechanism – A Calendar Computer from ca. 80 BC* (New York, 1975), 20.)

De Solla Price argues from the complexity of the Antikythera mechanism that Heron is 'on a side track from the main line of gearing' as it was developing from Ktesibios (*c.*300 BC) to the Middle Ages. None of the surviving ancient technical works, according to De Solla Price, records anything on a par with the Antikythera device. However, Heron understood the principle of differential gears, as shown here and at chapters 18 and 25, and in *Mech* 2.21 and the *Baroulkos,* where he proposes (albeit only theoretically) to use gear wheels rather than pulleys and axles (see Drachmann, 1963, 201–2). In the *Automatopoietikes,* other considerations may have made cords and pulleys preferable to gearing. Metal gears would be too heavy for an object of the size suggested in **3** 1. Wooden gears would probably be too delicate or wear out too easily.

Wooden gearing was used in antiquity for large-scale mechanisms: cf. the water-driven mill-wheel of Vitruvius 10.5.2 (see also Oleson, 'The Spring House Complex', in Anna Marguerite McCann *et al., The Roman Port and Fishery of Cosa,* 98–128, Figs V 1–76, XIII 1–7, Maps 8–14, for an archaeological reconstruction of a similar device).

7. The Greek word is ὄργανον, literally 'tool'. Most of the cords are wound around axles, pegs on axles, or drums on axles, in order to make the axles revolve. In chapter 13, a cord turns a stopcock to open a valve, and in **15** 4 one moves a hook to open a trapdoor; hence the use in the Greek of a fairly general term.

8. Heron is concerned, in both sections of the text, with improving on his predecessors and being up to date; his main concern is the mechanical aspect of automaton building, the mechanical principals involved, but he also wants the thing to look good to the audience.

9. As Schmidt points out, four would make more sense, since the palm as unit of measure was four fingers, but the text has ε', five. What Heron does not tell us is how wide the architrave, and therefore the covering over it, is to be; and this oversight in turn leaves us wondering about the size of the temple, the altars, the Maenads and the Dionysos. See

illustration for speculative dimensions. The diameter of the round (not oval) architrave must be at least 6 palms, to match the length of the long side of the base.

10. No such thing has been stated before. Schöne wanted to emend εἴρηται to εἰρήσεται since this arrangement will be described further in chapter 13. As for the raised surface, the word in the text, ἐντεταμένην, means 'stretched out, taut', and makes no sense. In his introduction, Schmidt suggests ἀνατεταμένην, which could mean 'raised up', which would give some idea of the roof coming to a point, as one would expect from a cone. See Schmidt, intro., LI, on Fig. 82, and apparatus, 350, line 15.

11. The measurements mentioned in **3** 1 are small enough to prevent the suspicion of an operator sitting inside the automaton. If mobile automata were displayed before large public audiences they would have to be big enough to be seen at a distance (Athenaeus, *Deipnosophistae* V.198f, Loeb Classical Library, Athenaeus, vol. III, 400–3); perhaps Heron is envisioning automata for private entertainment.

12. The rectangular tube in which the counterweight descends has to be set on top of the base, in the very centre, between the four columns. This is the best place for the sake of balance and for the cords from the wheel base and the top portion of the automaton to be able to reach the counterweight (see Querfurth's illustration, Schmidt, 357, Fig. 83b).

13. Deities: the text has δαιμόνων. The particular display Heron is describing involves Dionysos, so this could be what he means to say. Other conjectures are possible: Schmidt, 360, apparatus, line 6.

14. AH seems to turn out to be, not the length of the axle, but the length of that section of the axle which is between the wheels, or, more simply, the distance between the wheels. The axle itself is TY. See illustration.

15. Heron elaborates in mathematical terms on chapter 7. In practical terms, he is talking about making the inside wheel smaller than the outside one. This would result in an automaton which could *only* travel in circles. A more versatile arrangement is described in chapter 11.

16. The third wheel (ΥΦ) has to be on a separate axle.

17. 'Bar-frame' for κανόνιον in the text: Schmidt translates '*Latte*'. A slat, with a cutting (ἐκκοπή) in it. The *kanonion* serves as a frame for the wheel, but if it moves up and down in a dovetail on the side of the base, the wheel must either protrude through the side of the base, or the thickness of the *kanonion* must be greater than the radius of the wheel, or the *kanonion* must include a bracket projecting into the base to hold the wheel. Querfurth's drawing shows the bar-frame in a dovetail presumably up against the side of the base, with the wheel in a projecting bracket (Schmidt, 371, Fig. 90a).

18. Equal to the other axle.

19. That is, the base will move in a straight line in the direction determined by the angle of the turn.

20. Schmidt believes that this section is the work of an interpolator. He does not think that ΚΛ needs an axle-casing, since it is described in **11** 3 as simply rolling with the other wheels. This passage is inconsistent with the earlier one, but Heron begins it with the words 'But it is better ' (βέλτιον δὲ καὶ), indicating perhaps a contrast with what he has described earlier. This could be an alternate system – if you want to have more control over the third wheel on the base, set it up the same way as the other two, with its own string pulling it (see **11** 10).

21. The text reads γεγονέτω γὰρ τὸ πλινθίον καθ' ὃν τρόπον. The alternate reading, γενέσθω, is found in the manuscripts, and seems more logical. Schmidt's emendation, ἐν τῷ ἔμπροσθεν μέρει after the γὰρ, and τοῦ πλινθίουῷ for τὸ πλινθίον, is probably not necessary.

22. Schmidt finds this section suspicious (381, nn. 1–2). He expects the third wheel, ΚΛ, to be at the back of the πλινθίον rather than at the front; and if the cord is διπλῆ so that the wheel is in the middle of the axle (this must be what μέσον λαβεῖν and μεσολαβοῦσα mean) then the two wheels are closer together than if they were both at the outside, so the whole base is more likely to tip over. Moreover, he says in his introduction that using a double cord here would contradict Heron's earlier statement that the forward and backward motion of a single axle is controlled by a single cord (**6** 1, Schmidt, 358–9); and that it is odd that here, suddenly, the third wheel, previously described as being dragged along with the other ones,

is pulled by its own cord. Section **11** 7, along with this section, perhaps describes an alternate system to the first one Heron describes. See note 43 on **11** 7.

23. How big is the altar? If the platform is 6 palms wide, it might be $\frac{1}{2}$ to 1 palm wide. $\frac{1}{2} \times \frac{1}{2} \times \frac{1}{2}$ palms does not leave a huge space for the brazier or grate, but perhaps it is enough. The altar could not be airtight, or the fire would go out before the lid was opened.

24. Heron does not tell us what the fuel for the fire was either, or how to keep a very small amount of whatever it was burning for however long it took the display to reach that point. There is no provision for closing lid ZH, so the fire either burns out by the time the wood shavings are consumed or smoulders in the grate without being obtrusive. The actual lighting of the fires must be done by hand, as part of the process of setting up the automaton. See **28** 4, where the same sort of thing is being described.

25. Heron does not say how cylinder ΓΔ is attached to the rest of the automaton. Querfurth has it resting on an axle, which is presumably anchored in a bearing attached to or resting on the bottom of the platform on which the temple rests. The cylinder need be no longer than that to do its job; and it cannot be any longer than 1 palm, the thickness of the platform, plus the thickness of the temple floor, and the base AB on which the Dionysos figure stands.

26. The word in the Greek in πυρήν, 'knob, round projection'. Schmidt translates it '*Dach*', 'roof'. If there is a round knob on top of the shrine (at the peak of the roof?) it is probably an ornament, or a base on which the Nike can stand. It must be the roof itself which contains NΞ as well as axle ς'Z' and pulley H'. Querfurth's illustration (Schmidt, 387) has the Nike standing on a spherical projection on the apex of the roof, but labels the roof itself KΛM.

27. Reading περιφέρεια: 'the circumference of a half circle is traversed'. The thing described as moving 180° could be either the stopcock or the Dionysos: the statement applies to both of them.

28. Schmidt suggests μέρη for βάρη, and translates it 'to the other side' (*der anderen Seite*) (389). He also brackets the sentence and says he is inclined to leave it out, perhaps because there seems to be only one cord attached to the stopcock at **13** 5.

29. **13** 8–9. This second counterweight, M_B, is released by the trigger, which is itself released, presumably, by the action of the main counterweight. If M_B is the second counterweight in the two-counterweight system described in chapter 19, then the stopcock is also connected to it, as are the altars. However, both weights in chapter 19 are set in motion, not by a catch attached to a ring, but by the opening of the bottom of the container so that the millet seeds can run out.

30. The milk and wine must be released for the first time, before the Dionysos moves at all. If the Dionysos moved first, the cylinder ΓΔ could serve as a stopcock itself.

31. This almost sounds as if the balls are falling out of the bottom of the wheel base, which would not make sense. Maybe the box with the lead balls is really in the pedestal – inside one of the columns – rather than in the wheel base.

32. The same device, minus the cymbal, is mentioned in passing in **20** 4, as the thunder sound effect for the miniature theatre. (See note there.)

33. θωράκιον is rendered 'frame', based on Schmidt's '*Rahmen*' (391).

34. When the boards are held in place by the hook, they cover (actually, support) the wreath in the *thorakion*. I think ἐπιπωμασθῇ (line 16) refers to this action rather than to the boards, or the hinges, being covered in the sense of hidden, out of sight. Couture's Latin (Thévenot, 259) has *assulae . . . clausae remaneant*.

35. Schmidt wants to cut this phrase from the text, on the grounds that it is repeated a line later, and therefore may have been copied here by mistake. On the other hand, his app. crit. does not say anything about variant readings in the manuscripts, so it is possible that Heron may be emphasizing the importance of having the cord all the way in the groove, i.e. tight around the felloe, because the cord, as it turns out, is what moves the felloe. Schmidt is probably right that the idea could have been expressed without such repetition; however, lines 4–5 need something to explain where the cord is wound, if εἰς τὸ βάθος τοῦ σωλῆνος is stricken.

36. The axle isn't supposed to turn independently of the drum, since it is being used to wind the cord around the drum, so he probably means that it is spinning on its axle pins, which are stuck into something else.

37. That putting the Maenads on the felloe is what Heron has in mind is pretty clear throughout the passage, but he only mentions this at the end.

38. The lower pedestal (κάτω βάσεως) is the base with the wheels.

39. To increase the distance the base travels – without increasing the height of the tube and the distance of the counterweight's descent – Heron suggests decreasing the circumference of the axle, or increasing the circumference of the wheel. He prefers increasing the circumference of the wheel, but does not say why. In chapter 18, he goes on to a more mechanically interesting approach. The maximum diameter of the wheels depends on the size of the base (**3** 1), which hides them except for the small portion at the bottom (see **2** 7).

40. See app. crit. for line 6, p. 400; n. 3, p. 399; p. lvii, s.v. Fig. 98; Drachmann, 1963, 61–3, also 50–1. In Heron's discussion of the principles of leverage in *Mechanika* **2** 7–8 (Drachmann, 1963, 61–3), and, Schmidt says, in the ἐν τοῖς μοχλικοῖς of Philon of Byzantium, the arms of a lever are seen as the radii of two unequal concentric circles; thus the lever works on the same principle as the winch. A heavier counterweight is needed because a smaller axle equals a shorter lever-arm, which requires greater force to move the lever. *Mech.* **2** 7 discusses the relation of weight (force) to length of lever-arm, as does Archimedes, *On Equilibrium*, **1** 6–7, cited by Heron. Cf. also Aristotle, *Mechanical Problems*, 850a35–b11.

41. Schmidt adds the μή. It makes sense – Heron does use a slackness in the string to absorb the initial jolt at **9** 6.

42. It seems that each counterweight is to control the opening of the other's lid. Heron does not say exactly when or how hole K is closed: it must be either at the same time hole L is opened (perhaps by the same cord?); or after hole L is opened, by a cord activated by the descent of the second counterweight. The action of the second counterweight subsequently opens K again. This presumably would be the last thing that happens before the millet runs out in ΗΘΓΔ.

43. The ὥσπερ, 'as though', makes it sound as though Heron intends something other than cords to make the Athena move on the hinge; but he never explains what. At **29** 2, however, he describes the manipulation of this figure by two cords, but in **29** 1 she is mounted on a base, which seems to be an alternative to the hinge. Heron has the figure lying horizontally, apparently under the stage, with its feet attached to a hinge so that it can be raised and lowered from the hinge. He never explains how it will be made to turn on the hinge, and his language seems to rule out cords, although perhaps he means that the cords will not act on the figure directly, but only through the hinge, when he says she will 'appear, upright, *as though* drawn up by means of a cord'. At **29** 1, Athena appears on a base which has 'pegs at appropriate points'. Heron does not explain what the pegs are for. In **29** 2, the figure is pulled upright, and then apparently around in a circle (perhaps just back and forth, but with a turning motion?) and back down out of sight. Schmidt's illustrator, H. Querfurth, devised a plausible reconstruction of how this could have worked, with pegs attached to the bottom of Athena's base (W. Schmidt (ed.), *Heronis Alexandrini opera*, vol. I, introduction, LXIII–LXIX).

44. τῶν στατῶν is Schmidt's suggestion for filling an ellipsis here.

45. That is, different productions on stage work with the same devices backstage.

46. Schmidt, following Diels, fills the ellipsis here with ἡ δὲ 'Αθήνα ἐπὶη: Prou and R. Schöne read μηχανή – the crane was lifted, etc.

47. **22** 3–6. The closing and openings of the doors divide the show into five parts. Prou observed that the scene changes (marked by closing and opening of the doors) in Heron's Nauplios play have the apparent effect of dividing the drama into five acts. H. Weil (*Journal des savants*, 1882) argued, in response to Prou, that the five-part division was the result of technical rather than dramatic considerations. W. Beare, *The Roman Stage* (Northampton, 1964), contends that the five-act convention did not really exist in Greek or Roman drama, but was a pedantic invention of Evanthius and Donatus with help from ambiguous remarks by Horace and Varro. Beare therefore agrees with Weil that Heron's divisions were technically motivated, and that the automaton itself has very little relevance to theatrical concerns. The limitations of the puppet theatre differ from those of the full-sized stage: each medium adapts to what it does best. The story of Nauplios was the subject of a (lost) tragedy by Sophocles.

48. Prou takes the α with an overstrike, which appears here twice in the manuscripts, to be the letter α, designating first the loop, then the knob. Schmidt takes it as an abbreviation for 'first'.

49. That is, this mechanism (like all the others) is hidden from view?

50. <show> Schmidt, ὑποδεικτέον to complete the sense. how: reading πῶς with Schmidt, instead of πως with the manuscripts.

51. Prou, 65–6, suggests that the ὕσπληγξ here gets its name by analogy with that used in catapults, but that the use of the counterweight to control it was more practical for Heron's purpose than torsion power.

52. This device will work for the figures that are supposed to be using hammers and axes. Heron does not suggest a mechanism for sawing motion. We will see in chapter 25 that Heron envisions the 12 figures referred to in the overview (chapter 22, esp. 3–4) as being painted on the back wall of the theatre, except for the moving arms described here, so that they can be covered by the unrolling of the cloth painted with the scene of the men launching their ships.

53. τὴν πλευρὰν τοῦ πλινθίου, 'the side of the box', actually refers to the ceiling – see **26**. 3, where πλευρὰν has to refer to the top side. Here also, the top side makes more sense – but see illustration.

54. Reading παρὰ τοῦτο; Schmidt, 430, app. line 8.

55. Schmidt suggests inserting the negative, but would prefer to strike the passage.

56. Schmidt, 436, app. crit. line 4.

57. <Anchoring>, Schmidt's emendation, 438, app. crit. line 11.

58. Athena first appears painted on a cloth with Nauplios. In the next scene, after this cloth is hidden by the one showing Ajax swimming, a puppet Athena is pulled onstage, somehow: see notes 43 and 61.

59. As in chapter 12, there is no automatic mechanism to light the fire; you have to do it by hand.

60. The box containing the lamp is just like the altar in chapter 12, although the lid here is described as triangular (the shape was not specified in chapter 12), and here the flame is seen through a window in the pediment, and the box itself is hidden.

61. Thévenot's Greek text (273, line 8) is different from Schmidt's. Schmidt's version of the text has a second cord pulled, from in front of the hip. Thévenot's has one cord, from behind the hip, which pulls Athena upright and pulls her back down again. See note 43 on **20**. 2. No one seems to know what Heron meant this arrangement to look like. Querfurth devised an ingenious and complicated method of moving a cut-out figure of Athena around an ellipse at the front of the stage (Schmidt, LXIII–LXIX); it required the invention of many more details than Heron gives us.

62. The pegs to which the gut strings are attached are fixed in the ceiling of the box. It is not clear why there should be, below this, another ceiling which needs a slot cut into it. Whatever machinery is set above the stage should be hidden from the audience's view by the lintel.

63. This pin holds the cord which supports the board and the drop cloth, as in chapter 25. It must be connected via cord and pulley(s) to the counterweight.

64. <μύθοις>, Schmidt.

Acknowledgements

I am deeply indebted to Professor Peter Green, both for pointing me towards Heron of Alexandria and for guiding me through this project, which began as an MA report for the graduate school of the University of Texas at Austin. Professor Lesley Dean-Jones also supplied valuable insights, as did the anonymous reader for this journal. Of course, any mistakes are no one's fault but mine.

Archimedes the Engineer[1.1]

D. L. SIMMS

Enter *Legion*:
Je tue Archimède le Grec
Je tue ses machines
Je tue ses écrits
Je brûle ses papiers, au nom de la paix et du genre humain,
Je brouille ses rondes.[1.2]

INTRODUCTION

Very little is known about Archimedes' life. He was almost certainly a native of Syracuse and he told us that his father was Pheidias the astronomer.[1.3] He was acquainted with, and possibly related to, Hieron and Gelon, the tyrants of Syracuse.[1.4] A relationship is not incompatible with stories that he was a poor man, since Hieron had seized, and not inherited, power. However, there are lapidary inscriptions in the Theatre of Syracuse recording members of Hieron's family.[1.5] Archimedes' name cannot be among them or the fact would have been proclaimed. Other negative evidence is that Archimedes is not listed among the fifteen guardians who were to guide Hieron's young grandson, Hieronymus, during his minority.[1.6]

The most probable place for Archimedes to have studied was the Museum at Alexandria; this explains his knowing at least one Alexandrian mathematician, Conon, personally.[1.7] Diodorus (fl. 60–30 BC) stated that Archimedes had visited Egypt.[1.8] He returned to Syracuse where he carried out his mathematical investigations and also acted 'as a kind of occasional Chief Scientist and Engineer' to Hieron.[1.9]

During Hieron's reign, Syracuse had been the ally of Rome. After his death, the pro-Carthaginian faction seized power and the Romans attacked the city.[1.10] Archimedes played a decisive part in repelling the first assault, but was killed by a Roman soldier two years later when the Romans stormed the city.[1.11] He died an old man; Tzetzes (*c.* 1110 to *c.* 1180) claimed that he was born in about 287 BC, but to believe this depends on the credence that is to be placed on Tzetzes' tales.[1.12] His killing was against Marcellus' orders, although that may just be a gloss to salve a Roman conscience.[1.13] Marcellus did allow Archimedes to be buried in a tomb with a symbol that was specially designed to demonstrate one of his particular mathematical discoveries. Archimedes had requested his kinsmen and friends to place a figure above his tomb consisting of a cylinder circumscribing a sphere together with an inscription giving the proportion by which the circumscribing solid exceeds the circumscribed one. Cicero (106–43 BC) recorded that he searched for, recognized immediately and restored the tomb when he was quaestor of Sicily, 45 BC.[1.14] What is most remarkable, though this has been overlooked, is that, after the appalling destruction that followed the Romans' breaching the walls, at least one person was left in Syracuse who understood the importance of that theorem and was able to persuade others of its importance too. We do not know what Archimedes looked like; all existing portraits are imaginary.[1.15]

Archimedes' distinction as the greatest mathematician and physicist of antiquity has been unchallenged for centuries. His achievements as an engineer have proved more vulnerable, although they were the basis of his reputation for a very long period. Unfortunately, the numerous inventions and achievements that have been attributed to Archimedes, both in Classical times and since, were not recorded by him, or his records have been lost. The more reasonable ones include: a compound pulley, variously called a *polyspaston* or *trispaston*, the first multi-sheaved, the latter with three sheaves; a windlass; an endless screw; and a kind of balance or steelyard. Others are simply preposterous: devising the acoustic properties of the quarry in Syracuse called the Ear of Dionysius; the corvus (a beaked device for locking the attacking ship to another, so allowing soldiers to storm across); artillery, gunpowder and the steam cannon.[1.16] This latter group appears to be a consequence of the temptation, in the absence of other names, to make Archimedes the inventor of last resort. The second part of the present paper therefore examines whether some of the civil inventions or achievements attributed to him are indeed his: launching the Great Ship; moving the earth; the planetarium and the water-clock; and the water-screw. The third part assesses his contribution to the arts of war and the fourth considers his attitude to his engineering achievements.

LAUNCHING THE GREAT SHIP

According to Plutarch (*c.* AD 46 to *c.* 127), Archimedes was commanded to launch a ship because he wrote to Hieron that with any given force it was possible to move any given weight and declared that, if he were given another world to stand on, he could move the earth. Hieron invited him to put his theorem into practice and show him some weight moved by a tiny force. Archimedes chose for his demonstration a three-masted merchantman of the royal fleet, which had been hauled ashore with immense labour by a large gang of men, and he proceeded to have the ship loaded with her usual freight and embarked a large number of passengers. He then seated himself some distance away and without using noticeable effort, but merely exerting traction with his hand through a complex system of pulleys (*polyspaston*), he drew the vessel towards him with as smooth and even a motion as if she were gliding through the water.[2.1]

However, according to Athanaios (early third century AD?) Hieron instructed a Corinthian architect named Archias to build a great ship, the *Siracusia*: 'the wood was enough ... for the building of sixty quadriremes ... One-half ... of the ship he finished in six months ... this part of the ship was ordered to be launched in the sea, that it might receive the finishing touches there ... Archimedes the mechanician alone was able to launch it with the aid of a few persons ... by the construction of the windlass he was able to launch a ship of so great proportions in the water. Archimedes was the first to invent the construction of the windlass.'[2.2]

There are a few other references, one or two adding a little extra detail. Silius Italicus (AD 26–101?) only noted that Archimedes had moved ships.[2.3] Proclus (AD 410–485) reported that Archimedes made it possible for Hieron alone to move a three-masted vessel that Archimedes had built, intended for Ptolemy of Egypt, down to the shore.[2.4] Oribasios (fl. *c.* AD 450) stated that Archimedes used a *trispaston*, described as a tackle with two pulley blocks, each with three sheaves, to draw ships into the sea.[2.5] This has a ratio between effort and load of 1 : 6, which, as Dijksterhuis noted, would hardly be adequate to move a ship having a capacity of 50,000 *medimni* (Greek bushels), which Tzetzes related that Archimedes had hauled with his left hand.[2.6]

Most commentators have confined themselves to discussing what mechanical devices could have been adopted. What none of them notices is that the Greeks, like the Egyptians, were quite capable of shifting very heavy loads.[2.7] Thus, if the feat has any substance in fact, Archimedes must have done something quite exceptional.

There are a number of substantial difficulties in analysing this putative achievement in either of the circumstantial versions. The first is that the two are not merely in 'a slightly different form'; the details are contradictory.[2.8] Plutarch recorded that the ship was already part of Hieron's fleet, was not of exceptional size and had been beached by many men, and that Archimedes' device pulled it along the ground, whereas Athenaios related that the ship was most exceptionally large and the task was to launch it before the fitting-out stage began. Plutarch stated that Archimedes used pulleys, Athenaios that he used a windlass. Moreover, in Athenaios' account, Archimedes had assistants; in Plutarch's he did it by himself.

Drachmann, however, had no doubt that the feat was possible. He presented his arguments in two papers; in the first he was concerned to explain how the story that Archimedes had claimed that he could move the earth had arisen and in the second to prove that Archimedes had invented the crank.[2.9] Drachmann based himself upon Plutarch's version, on the grounds that Athenaios' version was much embroidered and confused with the building of the Great Ship.[2.10] He observed that Plutarch had not mentioned the distance that the ship moved, only that it was moved. He suggested that a few paces would be enough for a demonstration and believed a short distance made the story more credible.[2.11] Drachmann proposed that Archimedes' system combined a rope drum with a pulley and a series of endless screws. The rope was wound around a drum with a circumference of one metre, and the drum was turned by means of an endless screw engaging a wheel with 50 teeth so that the handle of the screw travelled one metre for each turn (of the drum). This gave a mechanical advantage of 50. Adding a pulley of five sheaves increased the advantage to 250. If the first screw were turned by means of a second with the same ratio, the advantage increased to 1 : 2500 and a third would raise it to 1 : 125,000. Drachmann recognized that some part of the power would be lost in friction, but thought that there would still be quite a lot left and implicitly assumed that the highest value was sufficient.[2.12] In the later version, the screw is used in combination with the *trispaston*, which Drachmann chose to interpret as a drum, wheels and ropes; this gave a ratio of 1: 250 (Figure 1). He now regarded this ratio as sufficient, but left unexplained the reduction from the earlier value. He added a crank handle to one end of the endless screw.[2.13] Other than the crank, his device had a general resemblance to the devices he found in Heron, though he did not note this.[2.14]

Drachmann thought that Archimedes had made this kind of calculation in advance of his boast and was telling Hieron about the enormous power latent in his new form of windlass. When Hieron suggested that there must be some limit, Archimedes replied: 'There is no limit. Just give me somewhere to stand and I shall move the earth.'

The first and minor reservation is that this is not the form in which Plutarch gave the quotation. He recorded that Hieron was deeply impressed. Hieron's dramatic response comes from a later source: 'In future we shall have to believe everything that Archimedes tells us.'[2.15]

The second reservation arises from Drachmann's interpretation of Plutarch's text. Plutarch said that Archimedes used a complex series of pulleys; he did not mention endless screws or Orabasios' version of the *trispaston*. No proper justification is given for producing a *mélange* of devices from two accounts, one of which he otherwise rejected as too embroidered.[2.16] The feasibility of operating endless screws in series is ignored; so is the immense difficulty of aligning the components of the

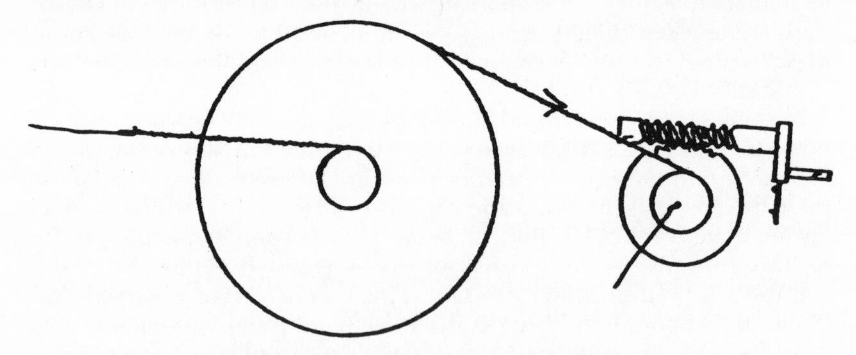

Figure 1 How Drachmann proposed to move the ship.

system accurately enough to ensure that the ropes in going through the system did not snag or snap and the teeth on the screws did not jam or strip.

The third reservation arises from the uncertainty about beaching practices. Herodotus recorded that 200 to 300 men could pull a ship of 75,000 kg on to a beach, though Moschion (a contemporary of Hieron) reported that one exceptionally large ship could not be hauled off by ten times as many men.[2.17] Plutarch's large merchantman could therefore have been dragged onshore by a large number of men, possibly divided into teams, each team with its own rope. However, while the ship was afloat, it was, by Archimedes' Principle, less heavy, and might have had some momentum of its own that should have assisted the men in their task of running it ashore. Once it was beached, a much greater effort, either in terms of numbers of men pulling or in operating a mechanical device, would be necessary to achieve the reverse movement in order to overcome the static friction between the ship and the ground. A considerably greater effort still would be required once freight had been loaded and passengers embarked. In addition, because static friction is greater than dynamic friction, a greater effort still is required to start to move Plutarch's ship than to continue to move it, and efforts to reduce the friction, by applying a lubricant to the ground around the ship, could only have a significant effect after the ship had begun to move.[2.18] Drachmann was therefore right to believe, though he nowhere made this particular point, that moving the ship a short distance would be sufficient to demonstrate Archimedes' boast, unless the effort exhausted Archimedes quickly, although Plutarch stressed the ease with which the feat was done.

A fourth, and related, reservation, not considered by Drachmann, is that the force had to be exerted away from the shore. In the Little Harbour of Syracuse, the pull could be exerted from the opposite beach, but only at the expense of requiring an impractically long rope; such a rope would be more likely to stretch and snap than actually tug. It is possible that the ship was beached near a headland, outside the main harbours, from which the system could be operated.

A fifth reservation is whether his system could transmit power. Drachmann had earlier concluded that gear wheels were only used for lifting burdens in theory, except for water-wheels where the power available was enormous or in the Antikythera mechanism where no great power was used – what Hill called *fine technology*.[2.19] That conclusion would, if correct, by itself be sufficient to destroy his argument about the mechanism for moving the ship. However, the hodometer described by Vitruvius, in the form reconstructed by Sleeswyk, did have gearing that transmitted a modest amount of power.[2.20] Lewis, too, has argued that the Romans had carriages that used gears that were close relatives of the hodometer. The fairly heavy-duty metal gearing required was paralleled by the surviving lock in the Temple of Romulus, although it, unlike the carriage gears, would not have to transmit much power.[2.21] Both inventions are attributed to Archimedes and accepting their claim means that

Archimedes did have heavy-duty gearing available to him, though whether it would have been strong enough to transmit sufficient power remains an open question.[2.22]

This leads to the sixth reservation, the efficiency of the system. The theoretical mechanical advantage of 12,500 may not be enough to move Plutarch's ship from its beaching, because of losses from internal and external friction. Although Drachmann knew that there would be some losses from internal friction, he blandly assumed that these would be overcome by the high mechanical advantage. Taking the weight of the loaded merchantman as 100,000 kg, and a coefficient of friction of wood on dry grass as 0.25, the force that has to be exerted by the operator is about 30 kg.[2.23] In high quality, five-sheaved pulleys, the loss in efficiency owing to friction is about 20 per cent and a reasonable value for one available to Archimedes could be double that.[2.24] The frictional losses within the screws would almost certainly have been much higher still, but, taking the same value, the efficiency would fall to a sixth. The force that has to be exerted by the operator thus rises to 180 kg; this is feasible, since the force that human power can exert for a few minutes turning a crank is about 160 N (kg m/s^2) and that exerted in turning a capstan is not that much lower.[2.25]

The next reservation is whether any rope could take the strain, another question that Drachmann did not consider. At one end, that of the operator, it could have been no more than 180 kg, but at the working end the rope has to take the strain of the ship, initially the equivalent of 400,000 kg. The strain might have been reduced were a number of these systems operated together, but Plutarch emphasized that Archimedes did the job by himself. Now Welsh suggested that the *hypozomata*, or hogging truss, the twisted rope cable holding the ends of the Athenian trireme together, had a diameter of 40 mm, and that it had to withstand a strain of 20,000 kg.[2.26] Assuming that the strength of the rope is proportional to the fourth power of its diameter, a rope with a diameter of 10 mm might be able to withstand a strain of about 80 kg. It is thus just feasible that the rope would take the strain imposed on it by the operator, but from these admittedly very crude calculations, the strain at the end attached to the ship would break any rope available to Archimedes and this conclusion is without prejudice to the other intrinsic defects of the system that Drachmann proposed.

An alternative is to replace the rope with an iron chain; Archimedes had run a chain over a pulley, either to release heavy weights upon or to operate a claw to grab the Roman ships.[2.27] The loss of power owing to the friction of the chain with a single pulley is relatively small, but in a five-pulley device it could well be too great for the pulley to operate. Friction would also be high in the drum, but probably less critical. Although the Greeks used wrought iron for the purposes of lifting and load-carrying heavy weights, there is no evidence that iron chains could take a load as high as 400,000 kg. Additionally, and finally, Drachmann's addition of a crank to his system is wholly unargued; even had he offered a substantial case, the windlass, credited to Archimedes by Athenaios, would perform almost as well.

Despite Drachmann's dismissal of the version by Athenaios, it is interesting to consider the feasibility of Archimedes launching the *Siracusia.* Part of the story raises no doubts. First, the dimensions of Athenaios' ship are compatible with those described elsewhere; Ptolemy IV Philopater built a *tesseraconteres* (40-oared ship), some 140 m long.[2.28]

Second, such large ships could be launched successfully from cradles and, later, dry docks, as Athenaios himself described, in an account taken from Callixeinos of Rhodes, third century BC.[2.29]

> At the beginning this was launched from a kind of cradle which, they say, was put together from the timbers of fifty five-bank ships ... Later ... a Phoenician conceived a method of launching by digging a trench under the ship equal in length to the ship ... This was in effect a dry dock.

While there is no suggestion of a dry dock in Athenaios' account, the ship might well have been built on a kind of a cradle resembling that described by Callixeinos. This system should be able to overcome any sagging stresses.[2.30]

As with Plutarch's ship, Archimedes would have had to exert his force *away from* the shore. With Athenaios' version, however, it is possible that, had the ship been constructed on a creek or cradle only a little way inland, the device could have been mounted between the ship and the water. Another difficulty is that Athenaios reported that an Archimedean screw was operated by one man to pump water out of the hold. Landels calculated that a single screw would not have been able to lift the water high enough to drain it off.[2.31]If the claim that a single screw was used is suspect, then that casts additional doubt on the story.

Archimedes' task was to overcome the friction between the ship's bottom and the cradle. In Athenaios' version, that could have been reduced by using a lubricant between the cradle and the ship.[2.32] None the less, the same problems arise as in Plutarch's version; in particular, the rope attached to the ship has to take an enormous strain. But since Archimedes had helpers and if each of them operated a similar device, each attached to the ship, the strain on the individual ropes would have been reduced, especially where the ropes were replaced by chains. It remains extremely dubious whether Drachmann's system, or anything similar, could have launched the Great Ship.

Drachmann did not therefore provide a practical device for moving Archimedes' ship and did not justify his claim that Archimedes invented the crank. Dijksterhuis's comment about the feat being fantastic remains to be refuted: that devices capable of delivering what Archimedes must have required of them only existed in the ideal realm of rational mechanics, in which friction and 'resistance of the air' [*sic*] are eliminated. His proposition that the story grew out of Archimedes having demonstrated to the Syracusans how to move a heavy weight by a set of devices providing high mechanical advantage has attractions.[2.33]

It is uncertain how far Archimedes was aware of the problems involved in the actions required to move a ship in the circumstances of either version. Athenaios attributed to him an interest in the fitting-out and launching, but not the design, of ships. None the less, two authorities have argued that Archimedes did consider ship design or building. Landels suggested that Archimedes' focus upon the paraboloid of revolution in *On Floating Bodies* reflected a practical interest in ship-building – the cross-section of a hull being approximately a parabola.[2.34] Tee agreed: Archimedes produced brilliant analyses of stability for simple geometric bodies.[2.35] Lloyd disagreed: he thought that Archimedes himself made no such connection and that his idealized mathematical study is as far away as could be from matters of practical concern.[2.36] Knorr noted that Eratosthenes specified that a solution to the problem of the duplication of the cube would prove valuable in the building of ships, catapults and other engines of war. He also noted that ancient authorities associated Archimedes with all three technologies, and suggested that, since Archimedes corresponded with Eratosthenes, he might have devised his method for this purpose. The claim is weakened by there being no evidence for this specific subject being part of their correspondence and no direct evidence that either of the solutions is by Archimedes.[2.37]

MOVING THE EARTH

The story that Archimedes boasted that he could move the earth has several different forms. The earliest is by Plutarch, who recorded that Archimedes wrote to Hieron that with any given force it was possible to move any given weight and 'declared that if he were given another world to stand on, he could move the earth . . . Hieron . . . invited him to put his theorem into practice . . . The chosen demonstration was moving a three-masted merchantman.'[3.1] Pappus, however, claimed that the boast arose from his solution to the problem *to move a given weight by a given force* in which Archimedes uttered the famous saying: 'Give me a place to stand on and I can move the earth.'[3.2]

Simplicius (first half sixth century AD) said that it derived from an Archimedean invention, a *charistion*, a steelyard, a weighing balance with unequal arms:

> When Archimedes made the instrument for weighing called *charistion* by the proportion between that which is moving, that which is moved, and the way travelled, then, as the proportion went on as far as it could go, he made the well-known boast: 'somewhere to stand and I shall move the earth'.[3.3]

Tzetzes' version, as might be expected, is a little more colourful: 'Archimedes is alleged to have said in the Dorian dialect of Syracuse: "a place to stand on! and I will move the entire earth with a *charistion*".'[3.4]

Drachmann scorned the idea of Archimedes, the practical mechanic who could lift Roman warships bodily out of the water, offering to move the earth either by a steelyard or by a simple lever. Since the word *charistion* is found in this connection in these two authors only, he thought it quite likely that Tzetzes has misunderstood Simplicius, both of them quoting the Greek in the Doric dialect. Simplicius actually reported that Archimedes invented the steelyard by employing what is now known as the Golden Rule of mechanics, not that Archimedes boasted of being able to move the earth using it. The boast only expressed the extreme application of that rule.[3.5]

Dijksterhuis in part agreed.[3.6] However, he argued that only the much later writers mention this instrument; Plutarch, Heron and Proclus did not. He suggested that Archimedes demonstrated the Golden Rule in public, stating it in the form the 'means for moving the earth by a given force from a fixed *point* outside the earth', and that his demonstration grew in the imagination of later generations into his moving the great ship.[3.7]

The use of the word *point* by Dijksterhuis in this particular comment, but not elsewhere, where the more usual and accurate translation *place* appears, is quite unsatisfactory. Archimedes could never have permitted the word *point* to have appeared in a manuscript. Had he done so, he would have committed the same offence that, in the *Sand-Reckoner*, he had criticized Aristarchus for; Archimedes objected that Aristarchus, in expounding his heliocentric theory, had given a point magnitude.[3.8] To operate a lever, a place to stand is necessary, not a point.

None the less, Hannah Arendt adopted the term 'an Archimedean point' as a metaphor for the (largely) Western way of life dominated by scientific thinking.[3.9] Although she appears to have started from the version by Pappus, an Archimedean standpoint is defined as one taken, wilfully and explicitly, outside the earth.[3.10] Her main chapter on the discovery and development of this way of thinking is headed by a quotation from Kafka.[3.11] 'He found the Archimedean point, but he used it against himself; it seems that he was permitted to find it only under this condition.' Arendt interpreted Kafka's remark as a warning not to apply the Archimedean point to man himself and to what he is doing on earth. If that is done, 'it at once becomes manifest that all his activities, watched from a sufficiently removed vantage point in the universe, would appear not as activities of any kind but as processes.'[3.12] However, with the coming of space exploration, her use of this metaphor has become obsolete. The astronauts who had to repair the satellite that relayed the Barcelona Olympics were able to do so because they, unlike Archimedes, had a lever and a *place* to stand outside the earth.[3.13]

THE PLANETARIUM OR SPHERE

One invention of Archimedes that is particularly well authenticated is the planetarium or sphere, though few details of its manufacture and operation survive. The book describing it, *On Sphere-Making*, was lost by the time of Pappus (fl. *c.* AD 350); he relied for its existence on the authority of Carpus of Antioch (date unknown).[4.1] Cicero actually saw two of the contrivances, those looted by Marcellus from Syracuse; one was retained by the family and the other, a much more elaborate version, was placed in the temple of Virtue.[4.2] Plutarch, in his third, and most fanciful, version of how Archimedes was killed, stated that it was one of the objects being carried by him to Marcellus.[4.3]

There are several other laudatory references to the extraordinary object, one well beyond the end of the Western Roman Empire, and all, with two exceptions, in Latin. Cicero mentioned them twice more.[4.4] Ovid at the beginning and Claudian at the end of the Principate introduced the device into their poems.[4.5] Other comments are in writers as disparate as the Christian apologist Lactantius (*c.* AD 250 to *c.* 320) and Firmicus Maternus (*c.* third century AD), also a Christian, but better known as an astrologer.[4.6] Macrobius (fl. *c.* AD 400) and Martianus Capella (fl. *c.* AD 500), the late Encyclopaedists, note its existence.[4.7] The last mention of the planetarium is in Cassiodorus (AD 487–583). Writing on behalf of Theodoric, King of the Ostrogothic Kingdom of Italy, he commanded Boethius (*c.* AD 480–524) to construct two clocks as presents for the King of the Burgundians; one of these appears to be a water-clock which showed the movements of the planets and the causes of eclipses. Cassiodorus did not attribute the invention to Archimedes. This is despite his reference in the letter to Archimedes the mechanician, whom, he stated, Boethius (in an otherwise unknown text) had translated into Latin for the Sicilians, and despite the implication that this text concerned the planetarium.[4.8] Sextus Empiricus (fl. AD 180) has the other reference in Greek.[4.9]

Not surprisingly, the references are contradictory about the materials from which they were made and there is little to be gleaned from any of them about how it worked. They agree that the planetarium was some kind of sphere constructed to imitate the motions of the sun, the moon and the five planets against the fixed stars in the heavens. Sextus' words suggest that the instrument was widespread. While he stated that it was made from wood, Cicero, who actually saw Archimedes' devices, had much earlier stated that those were made from bronze, Claudian said it was made of glass and Lactantius that it was made of brass.[4.10] Pappus made the only reference to its workings; he recorded that 'those who are versed in the making of a sphere can produce a model of the heavens by means of a uniform, circular motion of water.'[4.11] Presumably, this is why Hultsch conjectured that Archimedes' version worked by hydraulic pressure using a principle already utilized by Ktesibios (*c.* third century BC), though Pappus did not actually specify that the version by Archimedes did.[4.12]

Another source may be relevant, the short treatise *On the Construction of Water-Clocks* attributed by the Arabs to Archimedes. Much doubt has been expressed, as Hill recorded, that it be by Archimedes at all.[4.13] Nevertheless, Hill concluded that the treatise was fairly certainly based upon Hellenistic models and that at least a part of it was translated into Arabic from Greek.[4.14] He believed it right to give some credence to the unanimity of the Arab writers that the sections on water machinery and the release of balls were by Archimedes.[4.15] Thus Archimedes did invent the basic machinery and a single time-recording device and his description of it was incorporated in an extended treatise by Philon, who added the next two sections. The actual manuscript that Hill translated was completed by two more sections, one by a Byzantine craftsman and the other by an Arab worker. Hill offered a reconstruction of a water-clock using continuous machinery. The most recent commentator, Sleeswyk, modified the design to allow the gearing mechanism to operate intermittently. He also introduced a feedback mechanism that was derived from one of the water-clocks attributed by Vitruvius to Ktesibios. He also noted a passage in Procopius (b. *c.* AD 500) about a clock in Gaza that matched the description of the stones dropping out of the eyes to mark the time.[4.16] The Arabic water-clocks of Toledo, with their complex-gear trains, and the Chinese equatorial armillary spheres that rotated by water power, may both have descended from Archimedean planetaria.[4.17]

Acceptance of Hill's conjecture that the original work on water-clocks was by Archimedes confirms that Archimedes was capable of designing a comparatively complicated mechanism, worked by water. It would justify Hultsch's assumption that some types of sphere worked by water power, and to provide water power for a sphere ought to have been much easier and well within Archimedes' capacity.[4.18] Not all of them might have been so powered; Plutarch could not have supposed that Archimedes would have been carrying one operated by water power when he was killed. Cicero reported that two designs were brought to Rome. Since the later Antikythera mechanism, the portable sundial of Byzantine origin and the Mechanical Calendar of Al-Biruni were worked by hand, so might have been the earlier Archimedean sphere.[4.19] Only the recovery of the lost work on *Sphere-Making* could provide definitive answers to the questions: of what they were made and how they were operated.

ARCHIMEDEAN SCREW

The water-screw or cochlea is a device for lifting water short heights; its invention is attributed to Archimedes during his visit to Egypt.[5.1] Vitruvius, in the only description surviving from Antiquity, said that it consisted of an endless screw in the form of a spiral with eight windings fixed to the inside of a wooden cylindrical tube, sealed with pitch and bound with iron straps. When the cylinder is rotated, water travels up the screw and empties at the higher end.[5.2] The rotating force was supplied by the application of the feet of men to cleats on the outer shell in a treadle-type motion. Vitruvius also advised on the angle for its operation, which Landels calculated as 37 degrees. Landels commented that this offered the maximum possible lift at the cost of a reduced output, adding that the upper limit at which it will pump water was 45 degrees.[5.3] Modifications to the screw, each operated at its own specific, were made to meet local needs, largely depending upon whether output was more important than lift.[5.4]

Oleson compiled and reviewed a comprehensive set of literary, papyri and archaeological references to the pump.[5.5] He concluded that its original and main use was for transferring water out of streams or rivers on to fields for irrigation purposes.[5.6] For this it had notable advantages: a high rate of discharge at relatively low lift, a low susceptibility to clogging in turbid water, ease of installation and transport. Diodorus explained that the device threw out enormous quantities of water with little effort.[5.7] All these factors made it suitable for irrigation in Egypt and supported the Egyptian connection. The reign of Ptolemy II Philadelphus (283–246 BC), who was particularly concerned with the reform of Egyptian agriculture, coincided with probable dates of Archimedes' visit to Alexandria.

These properties made it valuable in other applications in ancient times.[5.8] Athenaios provided the only description of its being used for pumping water out of the hold of the *Siracusia.* However, Landels calculated that a single screw would not have been able to lift the water high enough to drain it off.[5.9] Neither he nor Oleson considers the possibility that the account by Athenaios is dubious on other grounds, so that the claim that a single screw was used may be equally dubious.

Oleson gave a full list of surviving attributions to Archimedes: Moschion, a contemporary of Hieron II, the source for Athenaios, Agartharcides (fl. 180–116 BC) and Poseidonius (fl. *c.* 135–51 BC); the last was the source for the attribution by Diodorus; finally, Bishop Eustathius of Thessalonika in *Commentarii ad Homeri Iliadem XII 293* in the twelfth century.[5.10] He also listed those of the first centuries BC and AD who did not mention Archimedes' name: Strabo, XVII 1, 30; Philo Judaeus, *De confusione linguarum* 38; Heron of Alexandria, *Dioptra* Chapter **6** 200 and ILS no. 8903.[5.11] He suggested that this group were concerned with the technical aspects of the screw rather than with giving an historical account. The difficulty with this proposition is that Vitruvius frequently gave the names of the inventors of the devices he described: admittedly he did not attribute any weapons of war to Archimedes, but he did attribute the force-pump to Ktesibios.[5.12] Heron (fl. AD 62–150), too, might have been expected to name Archimedes.

Vitruvius stated unequivocally that the screw was operated by the pressure of feet on a kind of treadmill.[5.13] This method is not unreasonable. A screw is difficult to turn by hand with a crank because the operator would be exposed to almost continuous wetting.[5.14] Using the feet in this manner is not easy since they can easily slip on the cleats, but this would punish the slave rather than waste water. Vitruvius' method is confirmed by a number of artefacts; for example, a negro working an Archimedean screw with his feet is shown in a Roman terracotta, dated after 30 BC.[5.15] A windlass or capstan is an alternative if attached below the top, always provided that the hands prevented the screw from rotating in reverse, but this should be easier to achieve with a capstan than with feet pushing against the wooden cleats.

The crank handle was not known in Classical times, though occasional claims have been made.[5.16] The screw illustrated in a manuscript of Taqi al-Din does not show how it operated, although the Arabs utilized the crank for some purposes.[5.17] The accepted view is that the crank did not appear in Western Europe until medieval times, though a representation of a crank handle has been recorded in China of the Han Dynasty (first century BC) and earlier descriptions have been found.[5.18] De Solla Price, although he thought the use would be remarkable, did suggest that the Antikythera mechanism might have had a crank handle to move it, an idea perhaps inspired by Drachmann.[5.19] White initially endorsed the view that the crank was a medieval device.[5.20] In his conclusions, however, he accepted an opinion derived from Healy that one of the Archimedean screws found *in situ* at Alcaracegos in Portugal was turned by a crank.[5.21] Healy's statement was unsupported by a specific reference, as was that by Ann Woods. Now Healy, White and Woods all reproduced an illustration of a screw from the Centenillo mine in Spain without a crank (which they attribute to Palmer).[5.22] In fact, Palmer's paper was confined to his personal observations of water-wheels in Roman mines. His paper was followed by oral communications, the fullest that by T. A. Rickard.[5.23] It was he who described the relic in the Centenillo mine. He concluded by expressing his admiration for the knowledge shown of laws of physics and some mastery of mathematical calculation. Rickard also made it clear that he was summarizing his earlier account, where

> he had discovered in the deepest of the Roman workings at Centenillo several Archimedean screws . . . Five were found but as many as twenty may have been in use [to raise the water to ground level]. The barrel as a whole is supposed to have been revolved by a slave, who applied his feet to cleats attached to the outside of the barrel about midway. This is the explanation given at the mine, but I venture to doubt it, because the mechanical advantage would be so small, and the operation would, it seems to me, soon cause the barrel to leak.[5.24]

Oleson, though he agreed that the absence of the crank is all the more surprising since its principle appears partly formed in the rotary quern of the second century BC, did not accept that the crank on the Portuguese screw was Classical; he had earlier drawn attention to the unsatisfactory nature of nineteenth-century unsupervised excavations.[5.25] Were a crank actually attached to the particular screw, he would assume that it had a post-Classical origin.[5.26] He also argued convincingly that the so-called crank and flywheel, attached to the two bucket-chain pumps on the Nemi ship, were the result of incorrect restoration.[5.27]

Despite what Oleson considered the remarkable clarity of its design, the distinctive terminology applied to it in antiquity and the explicit attribution to a single inventor, he was puzzled that the origins of the water-screw have continued to be discussed by modern scholars.[5.28] The challenge is unique, because the custom is to attribute inventions to Archimedes almost wildly, such as Petrarch crediting him with the invention of cannon.[5.29] The attribution was challenged before the end of the seventeenth century; in 1684, Claude Perrault thought that the screw was older than Archimedes.[5.30]

Drachmann realized that Vitruvius had described a more complicated version of the screw than that actually known to have been in use in Spain.[5.31] Drachmann noted that a water-screw found in Spanish mines had only a single spiral of copper instead of the eight wooden spirals in Vitruvius. He argued that the eight windings derived from Archimedes' original design, which had adapted a water-drum with eight compartments. Drachmann suggested that Archimedes had been investigating the properties of screws and spirals theoretically, had had the idea of using the screw to lift water and had simply modified the water-drum to incorporate the screw. It was left to later engineers to realize that the design was more elaborate than needed. From this, Drachmann argued that the design in Vitruvius is that of Archimedes and that its users converted the original into a more practical version.[5.32]

But when did these modifications occur? Given that Vitruvius was a practical engineer aiming to inform practitioners, these modifications may have occurred after his time. This would explain why there would be no archaeological evidence before the period of the empire.[5.33] However, Poseidonius reported that the screw was in use in Spanish mines in the second century BC.[5.34] It is more probable that Vitruvius had not seen the screw in operation since it was rare in mainland Italy.[5.35]

Although Drachmann's argument is highly persuasive and has been reinforced by Oleson, there is a difficulty about retaining the conventional dating of the invention with his invocation of Archimedes' work on spirals.[5.36] *On Spirals* has an introduction to Dositheus, who had succeeded Archimedes' friend Conon as his correspondent, and it recorded that Conon had been dead for many years.[5.37] Thus *On Spirals* was written years after Archimedes had returned from Alexandria to Syracuse. If Archimedes invented the device in Egypt, he would not then have been ready to convert his theoretical studies into a practical device.[5.38] On the other hand, Archimedes did record that he had taken a considerable time before publishing these proofs, so that his reference to Conon can be interpreted to mean that they had discussed spirals together in Alexandria, but that Conon had died before he could investigate the theorems that Archimedes now enclosed.[5.39]

Drachmann's analysis and conclusions also demonstrate the problems in converting Archimedes' design of a water-screw to a working piece of equipment, especially one that was adapted to three different uses: irrigation, drainage of mines and bilge-pumping. The much simplified surviving artefacts show that the designs of working devices were modified to meet local needs; a different angle of operation was chosen and the pitch changed depending upon the conditions. The materials of construction were also changed to meet different operating conditions. Since the profits from mining rich ores were greater than those from agriculture, this application warranted the replacement of wood by the much more expensive copper. Healy commented that they were efficient enough until men discovered the advantage of making the hall excavations big enough to take an animal.[5.40]

The existence of these variants also reinforces Drachmann's argument that Vitruvius was describing the original design. It follows that, if Vitruvius were quoting from a design, rather than from a working model, then he must have taken his description from a written text and this could well have been one of the works of Archimedes that Vitruvius claimed he had incorporated into his text.[5.41] In suggesting an unknown name from the 'school of' rather than Archimedes himself, Oleson may have been inhibited by too great a reliance on Plutarch's claim that Archimedes refused to write about his practical devices.[5.42]

The Archimedean screw is thus one of the only two surviving examples of a piece of apparatus which was designed by a theoretical engineer in the Classical period for civil purposes, the other being the force-pump designed by Ktesibios.[5.43] Moreover, working models of the screw and the force-pump differ from the theoretical designs and differed according to the use to which they were being put.[5.44] They provide the only examples in that period of the normal process of civil engineering development.

Its disappearance in the West at the fall of the Western Roman Empire surprised Oleson.[5.45] This may have been due to its rarity outside the Mediterranean littoral.[5.46] It was reintroduced into Morocco in Byzantine times and from there spread to Spain and thence back to Western Europe.[5.47] The earliest known drawing of an Archimedean screw, with a crank, in Western Europe is in Conrad Kyeser's treatise *Bellifortis* of 1405.[5.48] Its angle is greater than the maximum and the artist has not tried to represent a practical device. Leonardo's drawing of a double screw is not much better, and that in Fra Giocondo's edition is equally unrealistic.[5.49] When Archimedes' name, portrait and some of his attributed inventions were incorporated into the trade signs of London opticians from the late seventeenth century onwards, an Archimedean screw was placed upon the bill-heads of a firm of optical suppliers and remained there until Victorian times.[5.50]

The device with its design fundamentally unchanged continues in use in the Middle East to the present day. Some now have a crank handle attached to the higher end; unfortunately, the date of this modification is not recorded.[5.51] Landels found the use of a crank uncomfortable; today screw pumps in Egypt have two men seated on the ground, one holding the mounting, the other turning the screw with a handle; they are much less efficient than in Classical times.[5.52] Others operate without a crank.[5.53] Oleson noted that with the rising cost of fuel and spare parts for motors the screw is once again being worked by hand in the Middle and Near East, as appropriate technology.[5.54] It is safe to conclude that Archimedes, through his invention of the water-screw, made a magnificent contribution to the betterment of the lives of almost all its users.

ARCHIMEDES AND THE ARTS OF WAR

Introduction

Syracuse was not a typical Greek city-state. It held a substantial part of Eastern Sicily and was one of the wealthiest cities of the ancient world. Hieron II (269– 215 BC), its tyrant, after his defeat by Rome in 263 BC, during the First Punic War, judged that Roman prospects were greater than those of the Carthaginians and decided that he needed Roman protection to survive.[6.1] For practical purposes, therefore, for most of his very long reign, and thus for most of Archimedes' life, Syracuse was a client state of Rome. The Roman rulers appreciated the alliance, particularly when, at their request, Hieron sent them supplies, following the disastrous defeats inflicted upon the Roman armies by Hannibal during the early part of the Second Punic War.[6.2] None the less, Hieron must have recognized that Syracuse could not rely wholly upon the Roman alliance because he persuaded Archimedes to devise means of defending the city.[6.3]

There was a party in Syracuse that wanted to switch to the Carthaginians, with Gelon, Hieron's son, among them, but he died just before his father, in 215 BC. Hieron advised his successor, his grandson, to hold to the Romans.[6.4] Hieronymus ignored that advice and changed sides.[6.5] The Romans thereafter watched him as an enemy.[6.6] When he was assassinated in 214 BC, Syracuse fell into turmoil and coup followed coup; the pro-Roman party seized power but they were overthrown by the Carthaginian party. After their defeat by the Romans at Leontini in 213 BC, they were driven back to Syracuse where they massacred the pro-Roman faction. Once Roman attempts to persuade the new rulers to return to the old alliance were ignored, Marcellus, the Roman Consul, attacked the city. Nothing is known about Archimedes' attitude to this confusion and carnage. Had he been pro-Roman, he could have fled the city and joined them, as did many others. However, the frequent changes in allegiances did not affect his attitude to using his machines to defend the city.

Polybius (*c.* 203 to *c.* 120 BC) described Archimedes' activities at length; Livy (59 BC to 17 AD) followed Polybius closely and there is a substantial account in Plutarch.[6.7] Polybius is the nearest in time to the events, he made a practice of seeking out contemporary witnesses whenever he could and he understood the technology of war.[6.8] His main deficiency is that he, like all other authors whose works survive from the ancient world, other than those of Biton and Heron, never described the technical design of the period's artillery that Archimedes used.[6.9] His historical methods were far superior to Livy's. Plutarch, although he had some technical understanding, wrote centuries after Polybius. The analysis that follows is therefore largely based upon the account given by Polybius, amended slightly by Walbank and Marsden.[6.10]

Three aspects of the part Archimedes played need to be considered: first, what were the weapons he used; second, did he invent the weapons he deployed or did he adapt those already in use; third, what effect did he and his weapons have upon the siege? A fourth question is: did his contribution to military science have any effect upon subsequent military strategy and tactics?

The Attack on the City of Syracuse

The Romans may have been as confident of victory as Livy, and particularly Plutarch, would have us believe, but Consul Marcellus must have been much more aware of the difficulties of capturing what was, as Polybius' description makes clear, such an uncommonly well-protected fortress:

> [the] strength of the defences of Syracuse is due to the fact that the city wall extends in a circle along high ground with steeply overhanging crags, which are by no means easy to climb, except at certain definite points, even if the approach is uncontested.[6.11]

Besides, given that the two sides had been allies for so long, Marcellus might well have learned from Hieron himself that Hieron had been able to spare 50 catapults for Rhodes after an earthquake, so that Marcellus should have realized that the city had ample supplies.[6.12] Marcellus might even have expected that Syracuse had novel weapons; it was the city where the catapult or bellybow (*gastraphetes*) had been invented and first used more than a hundred years earlier.[6.13] He might have been told by Syracusans who had fled the city that Archimedes had made sections of the walls stronger; even had he not, he would have been able to see the defences for himself. What he can hardly have begun to realize was that Archimedes had conceived, designed, built and organized a comprehensive system of defence by sea and by land, with ample supplies of ammunition, able to cope with different forms of attack at all distances from the walls, by day and by night, once the besiegers came within the maximum range of ancient artillery; some weapons took him completely by surprise.[6.14] What Marcellus and his generals cannot have comprehended were the talents

> of Archimedes ... to foresee that in some cases the genius of one man is far more effective than superiority in numbers. This lesson they now learned by experience.[6.15]
>
> Accordingly Archimedes had constructed the defences of the city in such a way – both on the landward side and to repel any attack from the sea – that there was no need for the defenders to busy themselves with improvisations; instead they would have everything ready to hand, and could respond to any attack by the enemy with a countermove.[6.16]

When Marcellus attacked by sea, he found that Archimedes had stone-throwers able to damage his ships at long range, and when some of his ships slipped inside their range, so that the stone-throwers shot over them, there were others with proportionately shorter and shorter ranges. These weapons stopped the attacks by day. Marcellus then attacked by night. His ships succeeded in slipping past the first set of weapons; that is, they came inside the dead angle or shortest distance at which their missiles could be delivered.[6.17] Archimedes had prepared for this. The walls were pierced at short distances with a series of loopholes of the height of a man and of about a palm's breadth on the outer side. Archers and short-range arrow-firing catapults were stationed inside the wall opposite these loopholes and, by shooting through them, they disabled the soldiers on the decks. Finally, when the Romans tried to raise their *sambucae*, or giant scaling ladders mounted on pairs of quinqueremes, Archimedes had engines ready all along the city wall, but concealed behind them, which reared themselves above the wall, with their beams projecting far beyond the battlements. Some of them carried stones weighing as much as ten talents, equivalent to 370 kg, and others large lumps of lead. Whenever the *sambucae* approached, these beams were swung round on their axis, on the equivalent to a universal joint, and by means of a rope running through a pulley a release mechanism dropped the stones or lead on the *sambucae*, smashing them and imperilling the ship and those on board. Other machines were directed against those advancing under cover of shields and thus protected from any missiles shot through the wall. These could discharge stones large enough to chase the assailants from the prow of a ship. This done, other machines could let down a grappling-iron attached to a chain, which could seize the prow of the ship. The man in control of the grappling-iron then lowered and made fast the other end of the beam inside the wall, so making the ship stand upright on its stern. The chain was then let loose by means of a rope and pulley, suddenly releasing the prow, whereupon some of the vessels fell on their sides, some capsized but the greater number went under water and filled.[6.18]

Polybius gave a similar account of the assault by land. Men at a distance were mown down by the shots from the stone-throwers and catapults. Those that did get near the walls were checked by the volleys of arrows from the loopholes or, if they attacked under cover of shields, they were destroyed by the stones and beams dropped upon their heads. The grappling-irons were used to lift up and then drop men and armour.[6.19]

Archimedes' devices were sufficient, not only to drive off all the Roman assaults, but also, by demoralizing the attackers, to force Marcellus to confine himself to a siege. It was two years before his troops successfully stormed the city in 212 BC.

> So true it is that the genius of one man can become an immense, almost a miraculous asset, if it is properly applied to certain problems ... The Romans ... had every hope of capturing the city ... if one old man ... could have been removed; but so long as he was present, they did not dare even to attempt an attack.[6.20]

In short, Archimedes had overridden Philon of Byzantium's (second century BC) conclusion that, after describing how to counter every weapon and every method of attack, besiegers normally held the advantage over the besieged.[6.21] Later, Josephus (b. AD 37/38) was to claim that the walls of Jerusalem would have withstood the siege had not the internal dissensions weakened the defenders.[6.22] The walls of Constantinople defied Philon's conclusion until an entirely new weapon had been invented.

Archimedes' Weapons

The devices may be grouped into four categories or classes, all usable against attacks by land and by sea:[6.23]

1. A range of stone-throwers and catapults designed to be capable of projecting heavy missiles over different distances.
2. Short-range arrow-firing catapults operated behind a set of narrow slits in the battlements, presumably shooting along fixed lines.
3. Beams swivelling upon a pivot, making a universal joint, with a release mechanism using a rope round a pulley at one end that could be fixed at the other and then released suddenly; these could drop stones and lead weights heavy enough to wreck a Roman *sambuca*, hole ships or devastate groups of soldiers protected by shields.[6.24]
4. Similar beams with large grappling-hooks operating from them, capable of seizing and overturning ships and grabbing men and armour.

The catapult bow appeared in 399 BC at Syracuse during the tyranny of Dionysius I.[6.25] It was made more powerful by the use of composite materials, but then became too heavy to be held by hand and had to be placed upon a pedestal. The flexible bow was replaced by a torsion spring by Alexander and his successors (*c.* 330–300 BC).[6.26] The largest could hurl a stone as much as 80 kg in weight, though lighter ones were normally used, with a maximum useful range of the order of 400 yards.[6.27] Marsden estimated that the large double-purpose engine mounted on the Great Ship could hurl a 3 talent shot or a 12 cubit bolt 200 yards.[6.28] Schramm carried out experiments to determine the ranges of these instruments, and although any modern reconstruction can only give a rough idea of the distances, these were substantial.[6.29] Students at Reading University constructed a Greek-type catapult that hurled a 23 kg shot for 100 m.[6.30] Altogether, Marsden concluded that ancient pieces of artillery had a useful range and reasonable accuracy in the hands of skilled aimers, and could often achieve impressive results.[6.31] Soedel and Foley agreed.[6.32]

Experiments to determine the relation between the tension in the cords and distance of throw had been carried out in Alexandria by 270 BC and thus during, or just before, Archimedes arrived there.[6.33] While the formulae were not wholly straightforward, neither they nor their application would have presented any difficulties to Archimedes.[6.34] There is nothing unexpected, therefore, about his ability to obtain the varying ranges from his stone-throwers and scorpions. Big machines were normally placed in front of and at the base of the walls, and Philon recommended this. Marsden interpreted Plutarch thus and commented that, in most Greek walls still standing to their full height, there was no sign of casements specially for placing of artillery. He concluded that Archimedes had set his artillery to match Philon's position.[6.35] However, in some places at Syracuse, the sea washing the shore was deep enough for the ships to approach very close to the walls. Athanasius Kircher measured this distance in the seventeenth century as 10 m. He assumed that the site had not altered in 1800 years; this was a dubious assumption and one that Kircher, as a student of geology, should not have made and Knowles Middleton should not have accepted.[6.36] Some of the main artillery must therefore have been placed in the casements, despite Lawrence's argument that they must be mounted below the walls to ensure a low trajectory, which was only possible on the landward side.[6.37] Placing them on the walls would have severely restricted their line of fire, certainly if mounted well behind a narrow slit.[6.38] That limitation may have led Archimedes to introduce the other weapons.

In the second category, Marsden considered that the modifications to the range of the scorpions were practicable. The cords were shortened, thus increasing the impact of the missiles, a simple application of the calibration formula. They enabled the defenders to bring fire almost to point-blank range, well inside the range of the heavy artillery on the ramparts.[6.39] Walbank thought the slits were very close together, but the Tower of Assos in north-western Turkey, built in the second century BC, with slit windows for bolt-projecting catapults set halfway up and broader openings for (small) stone-projectors at the top, has them fairly well spaced.[6.40]

There is little reason to doubt the efficacy of the weapons in the third and fourth categories either. Cranes or grappling-irons had been used to drop weights on besiegers by land and sea before.[6.41] Moreover, the Greeks were thoroughly familiar with pulleys, cranes and treadmills and used them for such tasks as lifting stones for building works. Both categories are straightforward applications of the same principles. The additional flexibility of Archimedes' compound pulley and his grasp of the principle of the lever should have added to their efficiency.

Polybius did not specify the length of the beams used to drop the weights. Since, in some places, ships could approach to within about 10 m of the walls, beams would have to be at least that length to be directly over the ship.[6.42] The builders of catapults had the equivalent of a universal joint so that they could be moved into position as Polybius recorded.[6.43] The trigger mechanism for releasing the stones and lead weights needed to be operated separately from that controlling the movement of the 'universal joint', but the spring loading that would be necessary should be readily adaptable from the mechanism for releasing the arrows from large catapults.[6.44]

Again, with the weapons in the fourth category, the mechanism for grappling the man or the prow of the ship would need to be separately operated from the system for manoeuvring the crane into position, but the same kind of spring loading used in the third category would be applicable and, like it, also fairly readily adapted from the mechanism for releasing the arrows. Landels reasonably complained that Polybius had been over-rhetorical; ships need not be hoisted right out of the water, but just high enough to ship water, leaving the rest to the laws of hydrostatics.[6.45] Landels was left with the impression of 'a crane mounted on a tower slightly higher than the fortifications, with a boom on a swivel mounting which could tip up and down, and which, if not pivoted at the centre, might have been balanced with a counterweight.' A block and tackle could have been mounted at the far end of the boom with an anchor point at the base of the wall, giving purchase for a tug-of-war team pulling on the cable. This, he believed, would be consistent with Biton's *sambuca*, mobile cranes, and the accounts in Vitruvius.[6.46]

Even the noise that the missiles made may have had their effect in creating panic in Roman ranks.[6.47] Josephus wrote, when exposed to the Roman artillery at Jotapata, of the terrifying sound of the missiles, made more terrifying at that site by the echoes from the mountains.[6.48] If they terrified the defenders, they must have boosted the morale of the users. It was left for Londoners towards the end of World War II to discover the even more terrifying sound of silence, the moment when the engines of the flying bombs cut out.

The Legendary Weapon

To the historically attested practical weapons, Silius Italicus added firebrands thrown from a high tower and this addition, the use of fire, mentioned in passing by Lucian (*c.* AD 115? to *c.* 180) and Galen (*c.* AD 129 to 199?), became, in the Byzantine architect and engineer Anthemius (sixth century AD), the 'unanimous tradition' that Archimedes had used a burning mirror to set fire to the Roman ships, to which he added the surmise that the distance was a bowshot off.[6.49] In the twelfth century, the legend was noted by Eustathius of Thessalonica, and Zonaras (first half of the twelfth century) and Tzetzes gave the only circumstantial accounts of the use of burning mirrors that remain. Both claimed to have read their accounts in Diodorus of Sicily and Dio Cassius (*c.* AD 155 to *c.* 235) – whose works were still complete – and Tzetzes in works by Heron, Philon and Pappus as well, although no such passages can be found in any of the surviving works of these authors.[6.50] Gibbon was scathing about their reliability, though clearly neither invented the story.[6.51]

Although Greek Fire was not invented by Kallinokos (seventh century AD), more than a century after Anthemius, less powerful, but still effective, versions had been available for a long time. No one needed a burning mirror with supplies of brimstone and pitch available; their only drawback was that the wind might change direction too rapidly.[6.52] Had burning mirrors been a tolerably useful weapon, Anthemius would have been summoned by Justinian to design them for his wars. Tzetzes was writing at the time of the most devastating wound ever inflicted by the Christians of the West on their own interests, the Latin conquest of Byzantium. Then was the time to build them. The present analysis of the weapons deployed by Archimedes confirms that there is no historical or military evidence that Archimedes used a burning mirror.[6.53]

The earlier conclusion that a burning mirror would not work has been strengthened by the calculations of Mills and Clift.[6.54] Combining these two analyses permits a fuller consideration of the tales of Zonaras and Tzetzes. The most curious aspect of their accounts is that they both recorded that a burning mirror was the last weapon to be used. Although a mirror is more effective the nearer the ships are to the shore, the closer they were the easier it became for a Roman soldier to put them out of action. These mirrors were made up of segments, flat plates whose angles with respect to each other could be altered to change the focal length. A metal mirror would not be that difficult to disarrange by the stones from slings – the glass of the period had too low a reflectivity to be usable. Another puzzle lies in Tzetzes repeating Anthemius' surmise – which was nothing more than that – that the mirror was deployed only when the ships had been withdrawn by Marcellus to a bowshot off, whereas Zonaras recorded that the ships lay at anchor without suggesting that they had been moved away from the shore.[6.55] This is inconsistent with Polybius, who recorded that the ships only came close to the shore during the attack that began at night; Tzetzes' farrago referred to the noonday sun.[6.56] Even had the attack begun just before dawn, the battle would have had to have lasted several hours, well beyond the operational effectiveness of the ships!

Given that any test of their claims must necessarily be based upon their descriptions, accepting either account requires that the mirror was used on targets very close to the shore and, in Zonaras' version, probably well within the range of the two arrow-firing weapons being used alongside each other, the handbow with a range of 50 m and the scorpion of about 150–200 m; because both were being fired from narrow slits behind the battlements, their actual ranges were probably less.[6.57] On this interpretation of Zonaras' version, the ships might have been as close as 30, or even 20, metres from the mirror, whereas according to Tzetzes they could have been nearly twice as far away.

Mills and Croft calculated that at 60 m Archimedes would have needed more than 400 m^2 of mirrors to produce a feeble flame.[6.58] Taking either of the shorter distances increases the maximum possible irradiance at the focus quite markedly and the possibility of ignition occurring quite radically; this in turn increases the chances of ensuring continuing ignition a little. Alternatively, the number of mirrors required to produce ignition could be reduced by more than half, making their deployment less difficult. In neither case would the size of the flame started on the boat bear the slightest resemblance to the conflagrations described by Zonaras and Tzetzes. The mirror would be much more effective as an anti-personnel weapon;[6.59] devastatingly painful third-degree burns would be inflicted within a second or two, though no commentator referred to this option.

Moreover, supposing that Archimedes had decided to use a burning mirror as a weapon, the question arises as to why he did not use it against the land forces. He had used all his other weapons against attacks by both land and sea. It might well have been more effective against land targets, as troops move fairly slowly and may often be stationary, when they would be very vulnerable to burns; the *testudo* (tortoise-shield) would be very exposed. Since neither Tzetzes nor Zonaras mentions the attack by land, this is further evidence of the inadequacy of their purported summaries of the accounts of Dio Cassius and Diodorus. Finally, had Heron, Philon and Pappus commented on the legend, why did Tzetzes quote only from Anthemius and offer a muddled account of his work at that?

When Were the Weapons Prepared?

Since Hieron had died two years before the siege began, Archimedes must have had more than that time to prepare the defences at his tyrant's command.[6.60] Moreover, the spectacular effectiveness of his armaments requires that he had had time to train the Syracusans in their operation; that training need not have taken long, given their long experience in the use of artillery. There is no question of Archimedes having had to improvise the defences at short notice; Polybius, Livy and Plutarch made it clear that everything had been prepared in advance.[6.61]

How Original Was Archimedes' Approach?

For the catapults in the first category, he may have been the first to have designed a set that, operated together, would cover almost the entire distance from ship to shore, and from the extreme range of the catapult to near the land walls. Any gaps at close range were covered with those in the second category. The scientific formulae for setting ranges of the catapults appear to pre-date the siege.[6.62] Marsden did emphasize the novelty of the special arrow slits in the walls.[6.63] However, neither Philon, a generation later, who recommended that towers and walls should be channelled for artillery and have similar slits or windows, nor Heron, who also described them, attributed their introduction to Archimedes.[6.64] Similar devices to those in the third and fourth categories had already been described by Polybius elsewhere.[6.65]

> at Ambracia in Aetolia ... while the rams continued to batter the walls and the long sickle-shaped grapplers to drag down the battlements, the defenders of the city made efforts to counter-engineer them, dropping by means of cranes, leaden weights, stones, and stumps of trees on to the rams, and after catching the sickles with iron anchors dragging them inside the walls, so that the apparatus was smashed against the battlements and the sickles themselves came into their hands.

None of the surviving handbooks of war technology ascribes to Archimedes any contributions to military practice. These were written by his younger contemporary, Philon of Byzantium, and by the two most notable later technical authors, Heron of Alexandria and Vitruvius; both made frequent references to Archimedes, and Heron described similar devices. Heron's failure to mention Archimedes in the *Belopoiika* could have been due to that work not extending beyond the third century BC;

it may have been in Heron's edition of Ktesibios' work on artillery, while the *Cheiroballista* probably described the artillery of his own time.[6.66] Vitruvius drew his account of war machines and their inventors from one Diades, who may not have named Archimedes.[6.67] Finally, the late Roman author Vegetius (fl. *c.* AD 400) and the slightly earlier, anonymous author of *De rebus bellicis* made no mention of Archimedes.[6.68] Hacker, in his review of the part played by scientists and technologists in the development of catapults, did not find it necessary to include any advance due to Archimedes.[6.69] These omissions imply that either the inventions themselves were not that remarkable technically or they were absorbed into the arsenal of weaponry fairly rapidly.

The other reason may be that his real contribution was not generally recognized or understood by his contemporaries and successors. Marsden considered that Archimedes' ingenuity, with the financial support of Hieron, had turned Syracuse into a fortress of the first order, which embodied all the latest ideas for defence, though he added that some extra contrivances were invented by the great engineer himself.[6.70] Lawrence made a much more general claim: it was the systematic deployment of these weapons, together with the improvements to the walls, against all methods of attack that was new and warranted naming Archimedes as the inventor of operational research.[6.71] The views of Landels and White were more negative:

> The devices employed to destroy their [Roman] scaling ladders and the ships from which they were mounted called for no new inventions but only for slight, yet highly effective, variations of the pivoting crane.[6.72]

Thus Archimedes probably did little more than modify some of the existing weapons, though it would go hard, given our understanding of his knowledge of the principles of mechanics and his ability to apply those principles in practice, not to believe that his devices were more efficient and effective than those previously built. It is his comprehensive, systematic approach to the art of warfare that is unique, and it is only proper to recognize Polybius' insight in realizing this; he recorded that Archimedes had prepared by land and sea for all possibilities.[6.73]

Effects on Later Military Practice

Beyond delaying the fall of Syracuse, his devices had no effect on the outcome of the Second Punic War. Despite the alliance between Carthage and Syracuse, Hannibal never asked for Archimedes' help, though the Carthaginian lacked a siege train capable of attacking Rome. Possibly in the confusion that followed after Hieron's death, no one thought to proffer or ask for that aid. Had Hannibal done so, and had that aid been accepted, the war might have had a different ending or lasted longer. Hannibal might still not have sacked Rome, but the defenders of New Carthage and of Carthage itself might have held out longer had they adopted Archimedes' systematic approach. It was not as though the Carthaginians were strangers to novel methods of fighting: Hannibal employed elephants. Nor were they strangers to mechanical weapons: Scipio Africanus found two sizes of catapults and two of *ballistae* in New Carthage, as well as an immense number of large and small scorpions.[6.74] There is a curious contrast between Hannibal's attitude and that of Marcellus, who ordered that Archimedes be taken alive, even if only to obtain the services of the greatest scientific adviser who ever lived, and who paid him the respect of ensuring him a proper burial in a tomb with an Archimedean symbol.[6.75]

The Romans took their knowledge of catapults and other weapons from the Greeks, though by the first century AD they had begun to make improvements.[6.76] There are descriptions of the use of these weapons by Caesar in his accounts of his battles, and by Josephus in his version of the conquest of Palestine.[6.77] When, in the Civil War, Caesar attacked Massilia, once a Greek colony, and in as strong a natural defensive position as Syracuse itself, he faced a multitude of engines that threw heavy beams and used cranes to dislodge his six-storey tower.[6.78]

None the less, despite Archimedes' true role in the siege being recorded and abundantly recognized by Polybius – 'prepared by land and sea for all possibilities' – and despite Polybius being an active figure in Roman society in the circle of Scipio Aemilianus, and despite the Roman recognition of Archimedes' capacity as a military engineer, his comprehensive strategy was never borrowed, either in attack or in defence. This failure could well have been due to the absence of the necessity until it was too late.[6.79] For example, at Jotapata, although Vespasian and Titus in the conquest of Palestine had a range of artillery, they had no short-distance devices such as those Archimedes had provided at Syracuse. Consequently, once the Jewish defenders had succeeded in coming inside the range of the siege-engines, they were able to attack the Roman infantry.[6.80] When the Goths besieged Rome in AD 537, they attacked the Aurelian wall near Hadrian's Villa. By arriving by stealth, they got within close range. They broke cover and opened their assault with such suddenness that the defenders were unable to bring into play their throwers (*ballistae*), which could only hit targets at their own elevation. The Romans (strictly Byzantines under Belisarius) were few in number and very hard pressed, but Procopius does not use these reasons to explain why shorter-range *ballistae* were not to hand.[6.81]

Knowledge of the Siege and of Archimedes' Role in It

The part played by Archimedes in the siege must have been general knowledge in the Greek-speaking Eastern Empire, from Hellenistic to late Byzantine times, all derived directly or indirectly from the account by Polybius. There was also that in Plutarch, and presumably those in the lost sections of Diodorus Siculus and Dio Cassius. Both Zonaras and Tzetzes described Archimedes' weapons and the effects that they had.[6.82] Other notices of Archimedes' weapons are found in Athenaios, who referred to a *ballista* capable of throwing a stone weighing 80 kg and to a javelin 6 m in length being thrown 200 m, while Proclus noted Polybius' comments on Marcellus' reaction to Archimedes' achievements.[6.83]

There is also the evidence of Athenaios that Archimedes devised weapons for the Great Ship.[6.84]

His efforts were well attested in the Western Empire as well. In addition to the lengthy account in Livy, and a much more fanciful version in Silius Italicus, there were snippets in Pliny and Quintilian, both late first century, and in Ausonius, who lived towards the very end of the Empire.[6.85]

Knowledge in the West

The date at which Archimedes' feats at Syracuse became known to the medieval Latin West is unestablished. Intrinsically, the discovery should have caused little surprise: the weapons were not that dissimilar from those still in use. The onager was not replaced by the trebuchet until the Middle Ages.[6.86] Livy started to appear in catalogues of manuscripts in the twelfth century, but his works were rare even in Petrarch's day.[6.87] Petrarch was the first to write a full biographical sketch of Archimedes that drew upon several accounts, and described the siege and the weapons.[6.88] Petrarch, with wild disregard for his own soberer accounts, attributed the invention of artillery to Archimedes, and so did Valturius, while Tartaglia added the invention of gunpowder as well.[6.89] Leonardo added the steam cannon.[6.90] Valturius gave a full account based upon Plutarch.[6.91] From then on, almost every commentator on Archimedes described them. Gibbon recognized that they were effective, though improvable.[6.92]

It was left to the mid-twentieth century to cast doubt, not merely on the effectiveness, but also upon the very existence, of the weapons. Koyré commented that it is 'for the legendary invention of war machines that tradition glorifies him'.[6.93] Lawrence and Clagett also employed that word.[6.94] The implication is unwarranted unless 'legendary' is being employed in the sense of 'astounding', though this does not appear to be the meaning intended. Equally unwarranted is Dijksterhuis's lukewarm view that the accounts of 'ballistic instruments ... in spite of evident exaggeration, may well contain a core of truth'.[6.95] Equally puzzling is the use of the word '*extempore*' in relation to the casements by Marsden,[6.96] or Lawrence's belief that the defences were prepared at the last moment, despite his recognition that the design of parts of the walls was of such exceptionally high quality that he inclined to attribute the improvements to Archimedes. Such walls have to be built in advance.[6.97] Finley, followed by Green, cannot have studied the evidence.[6.98]

Given the account by one of the greatest of all historians, supported by two other similar accounts, there is no reason to doubt that Archimedes had devised and established a comprehensive system for the defence of Syracuse at least two years in advance of the assault. Nor is there any reason to doubt that they had a totally unexpected effect upon the outcome of the Roman assault. As a direct result, Marcellus had to wait two years before his troops could successfully storm the city. This is the most fully attested achievement of Archimedes the engineer. He had demonstrated that the application of the scientific method could change the face of war. Bertrand Russell found that, 'Ever since the time of

Archimedes war has been a science, and proficiency in science has been the main cause of victory.'[6.99] Fortunately or unfortunately, it was not until World War II that the first of Russell's propositions became largely true and the second rendered highly probable.

ARCHIMEDES' ATTITUDE TO ENGINEERING

Introduction

Archimedes' civil and military inventions and applications of his knowledge to practical problems appear to reveal a man willing to undertake these tasks with vigour. Yet when Plutarch recounted the part Archimedes played in the siege of Syracuse and described some of his technical achievements, he gave a version of Archimedes' attitude that was totally different:

> he was a man who possessed such exalted ideals, such profound spiritual vision, and such a wealth of scientific knowledge that, although his inventions had earned him a reputation for almost superhuman intellectual power, he would not deign to leave behind him any writings on his mechanical discoveries. He regarded the business of engineering, and indeed every art which ministers to the material needs of life, as an ignoble and sordid activity, and he concentrated his whole ambition exclusively upon those speculations whose beauty and subtlety are untainted by the claims of necessity.[7.1]

In an earlier passage, Plutarch used slightly less tendentious language:

> Archimedes did not regard his military inventions as of any importance, but merely as a by-product, which he occasionally pursued for his own amusement, of his serious work, namely the study of geometry. He had done so in the past because Hiero, the former ruler of Syracuse, had often pressed and finally persuaded him to divert his studies from the pursuit of abstract principles to the solution of practical problems, and to make his theories more intelligible to the majority of mankind by applying them through the medium of the senses to the needs of everyday life.[7.2]

This strain is continued by associating indirectly Archimedes' attitude with that of Plato:[7.3]

> Eudoxus and Archytas [were] the originators of the now *celebrated and highly prized art* [italics added] of mechanics. They used it with great ingenuity to illustrate geometrical theorems, and to support by means of mechanical demonstrations easily grasped by the senses propositions which are too intricate for proof by word or diagram ... Plato was indignant at these developments ... attacked both men for having corrupted and destroyed the ideal purity of geometry ... For this reason mechanics came to be separated from geometry, and as the subject was for a long time disregarded by philosophers, it took its place among the military arts.[7.4]

Plutarch's Influence

Until Plutarch's works became widely known in the West by the mid-fifteenth century, information about Archimedes mostly concerned his part in the siege of Syracuse.[7.5] Sarton argued that the first glimpse the West had of Archimedes was of a magnificent engineer of almost superhuman powers.[7.6] Even after Plutarch's account became familiar, Gille decided that the Renaissance engineers, such as Taccola, regarded themselves as the heirs of Archimedes.[7.7] As late as the end of the sixteenth century, Guido Ubaldo, Galileo's mentor and friend, attacked the notion that Archimedes disparaged mechanics as base and vile and material and did not deign to write of it.[7.8] None the less, Plutarch's views obtained early converts, such as Michel de Montaigne, and they have been accepted for centuries:

> the tale of that geometrician of Syracuse who was interrupted in his contemplations in order to put some of them into practical use in the defence of his country: he set about at once producing frightful inventions, surpassing human belief; yet he himself despised the work of his hands, thinking that he had compromised the dignity of his art, of which his inventions were but apprentice toys.[7.9]

Comments and Commentators

Although Plutarch's views are no longer unchallenged, some take them for granted and more remarkably still defer to them.[7.10] Oleson endorsed the view that the design of a water-screw in Vitruvius was that of Archimedes, but suggested that the source that Vitruvius drew upon was of the *school* of Archimedes, not Archimedes himself, though Vitruvius claimed that he had incorporated works by Archimedes into his text.[7.11] Hacker, in his survey of the contribution of engineers or mechanicians to Greek and Roman military technology, claimed that these men, including Heron of Alexandria, drew upon a Babylonian tradition quite different from that of Archimedes. His attempt to separate Archimedes from the Alexandrian school is wholly dependent on a belief in Plutarch, despite Archimedes having trained at Alexandria, with Heron normally regarded as one of his followers.[7.12] Green accepted Plutarch's views and developed them in a less and less restrained manner.[7.13] He followed Finley in asserting that Plutarch's views on Archimedes matched the *ethos* of the ruling class and its alleged malign influence upon innovation.[7.14]

Some agree with Plutarch on the basis that Archimedes did not write technical papers.[7.15] Others are more guarded. Dijksterhuis wrote: 'If Plutarch is to be believed', and 'geometry at play ... seems to have occupied his mind to a greater extent than he [Plutarch], ... with his Platonic contempt of technique, is prepared to admit.'[7.16] The third group is more severely critical of Plutarch's motives and suggests that, as a pious Platonist, he was trying to establish Archimedes as of the same school. They argue to the contrary that Archimedes was renowned for his inventions, in a period renowned for its inventiveness, that he was much praised for them by Hieron and that his military inventions had defeated the Roman army. Plutarch's version was invented or at least exaggerated; it had to be treated with the utmost scepticism.[7.17]

No one, however, has examined how far Plutarch's comments are internally consistent or match those of other Classical commentators on Archimedes or, for that matter, those on other technological innovators. The evidence will now be examined against: Plutarch's general reliability as a witness; Archimedes' own works; other commentators; Plutarch's purpose in writing about Archimedes; and the relevance to Archimedes of the *ethos* of the ruling class in that period.

Plutarch as a Witness

General

Belief in Plutarch's reliability has declined sharply from the level of trust of some Renaissance writers, notably Montaigne. Helmbold and O'Neil concluded that he did not verify all of his quotations because his memory was so prodigious, though strictly his remarks about Archimedes are not direct quotations.[7.18] Moles concluded that Plutarch's concern with historical truth, in the conventional sense, was intermittent. He distorted the truth within clearly defined limits: 'he did not invent or ... impute behaviour which is not true'.[7.19] By contrast, Scott-Kilvert noted, 'Plutarch's fondness for a semi-fictional rather than a factual treatment of history', and 'it is when Plutarch's imagination is most strongly engaged that it is most likely to supplant the facts'.[7.20] Although these comments are applied to his Shakespearean heroes, not to Archimedes, the latter certainly engaged Plutarch's imagination sufficiently to produce what appears to be a long digression. In equating Plutarch's purposes in writing history with those of Diodorus, Green intended no flattery to either.[7.21]

The only reasonable conclusion is that while Plutarch may have believed that his comments were those of Archimedes, unless corroborated, they are not sufficient by themselves to be accepted as an accurate reflection of Archimedes' own.

Archimedes' Achievements

Plutarch's descriptions of Hieron's demands on Archimedes to prepare weapons, the weapons themselves and the part they played in the siege differ only in relatively unimportant details from those of Polybius and Livy. His three versions of how Archimedes actually met his death may well be owing to there being no eyewitness account of the killing; the first two, like that of the burial, are consistent with others, notably those by Livy.[7.22] His description of the tomb matches that of Cicero.[7.23] His accounts of 'Moving the Earth' and 'Moving the Great Ship' do vary markedly from those of other commentators, but this may well be owing to different oral traditions. His analysis of Archimedes' theorems and procedures in reaching them has never been challenged, nor have his descriptions of the delights of mathematics.[7.24] The story of the bath, told elsewhere, is consistent with that of Vitruvius.[7.25] In short, almost all of Plutarch's relatively long descriptions of Archimedes' achievements when checked against those by other authors do not differ significantly and, where they do, the differences are explicable.

Thus far, given this record, his opinions on Archimedes' attitude to engineering and the practicalities of life cannot be rejected without additional analysis.

Archimedes and the Active Life

The descriptions of Archimedes' supreme powers of concentration and his complete indifference to his food and his surroundings, presented here and elsewhere, may be exaggerated, but Plutarch recognized the possibility of such exaggeration and they are not unlike those ascribed to Archimedes' peer, Newton.[7.26] His statement that Archimedes took no interest in public affairs would be a natural extension of this indifference, though it is not mentioned, and therefore not contradicted by, any other writer. Indeed, had Archimedes been a kinsman of Hieron, as Plutarch recorded, itself an unsupported statement, this indifference would explain his absence from the list of those who were to govern the city during Hieronymus' minority. That attitude may be true of his old age, though it would not be consistent with Plato's own political activism, particularly in Syracuse itself; nor, as White remarked, can it be readily accepted from the inventor of a water-screw, albeit a device of his youth, that affected the lives of so many people in his own lifetime.[7.27]

The Death of Archimedes

Plutarch's third version of Archimedes' death is much the most elaborate and quite unlike that by anyone else. It states that Archimedes was bringing a variety of instruments to Marcellus when he was met by some soldiers and they, under the impression he was carrying gold in them, slew him.[7.28] Why should a man depicted as totally unworldly be carrying scientific instruments to the conqueror of his city? It is unbelievable.

Weighing the Evidence in Archimedes' Work

None of Archimedes' works that are extant, and very few of those known to have been lost, described technical matters.[7.29] Since no one in Antiquity is known to have collected Archimedes' papers together,[7.30] those that did survive were presumably the ones sent to inform his mathematical friends at Alexandria of his discoveries, to demonstrate his achievements and, with *The Method*, to reveal how he arrived at some of them, in the hope of assisting his successors to advance the subject.[7.31]

This last work, as Smith realized immediately on its rediscovery this century, demonstrated Archimedes' indebtedness to the mechanical aspects of his subject.[7.32] He first investigated some mathematical problems by applying mechanical concepts, notably the law of the lever, and so discovered many of his theorems. Admittedly, he did not accept that the approach gave rigorous proofs, nor was it about mechanics, but it was thoroughly mechanically minded.[7.33]

Of the lost works, the book on *Sphere-Making* is recognized by most commentators to be an exception to Plutarch's claim.[7.34] One other work is mentioned in Classical literature, *On Balances.*[7.35] The Arabs allowed him a third, *Water-Clocks.*[7.36] A fourth could be the original source of Vitruvius' description of the screw.[7.37] In addition, Vitruvius has an *unequivocal* statement that Archimedes *wrote* about machines. He was nearer in time to Archimedes than Plutarch.[7.38] Johannes Tzetzes, a thousand years later than Plutarch, several times listed technical works by Archimedes.[7.39] Furthermore, there were good reasons for Archimedes not writing accounts of some of his techniques. Some were much better demonstrated and those that were for military purposes, devised at Hieron's command, could not sensibly be disclosed before they were used; afterwards it was too late.

Of course, if Sleeswyk is correct and Archimedes designed a hodometer for the Romans, then we have an additional device that Archimedes left undescribed, unless Vitruvius used him as a source without acknowledgement.[7.40] Again, if Lewis is correct, and Archimedes invented the south-pointing chariot, there is another lacuna.[7.41] These omissions hardly affect the issue since no one would expect Archimedes to write up all his inventions.

Given the evidence in Vitruvius and Tzetzes, the procedures adopted in *The Method* and the contents of some of the lost works, to accept Plutarch's claim that Archimedes disdained to write on technical matters would be to show a dangerous disregard for the evidence.

Plutarch, Plato and the Corruption of Geometry: Despising His Inventions?

Plutarch recorded that Eudoxus, Archytas and Menaechmus were attacked by Plato for using mechanical constructions.[7.42] However, his statement that Eudoxus and Archytas were the originators of the now 'celebrated and highly prized art of mechanics' is on their side;[7.43] it matches neither his avowal of Platonism nor Plato's views. Was he out of sympathy with Plato or was he just unclear? In any case, the statement neither claims nor proves that those were Archimedes' views. Far from agreeing with Plato, Archimedes had explicitly written *The Method* to explain his mechanical technique, which is an improvement on those of Eudoxus and Manaechmus. In addition, he recorded in that book his admiration for both Eudoxus and Democritus, singling out the latter for his achievement in obtaining a formula without actually proving it.[7.44] Democritus was totally unsympathetic to Platonism. It follows that, as Santillana recognized, Archimedes' rejection of Plato's precepts was just as strong as that of Eudoxus, Menaechmus and Archytas.[7.45]

Furthermore, other evidence suggests that Archimedes was no follower of Plato. Plato's name is associated with the relationship between the five regular solids and in particular that of the fifth solid, the dodecahedron, with the universe itself.[7.46] Plato, mainly in the *Timaeus*, but also in other dialogues, developed his views of an earth-centred universe, including the speeds and numerical estimates of the relative distances of the planets, though how these figures were obtained is unknown.[7.47] Later Platonists, according to Macrobius, gave different figures, which again, as Heath noted, 'have no basis in observation'.[7.48]

Archimedes, again according to Macrobius, determined the distances between the planets.[7.49] Chadwick commented that Macrobius 'was aware of the need to reconcile Platonic principles laid down in the *Timaeus* with the calculations of the mathematicians, especially Archimedes.'[7.50] The inference is that Archimedes, and other mathematicians, rejected Plato's figures. Someone who, in the *Sand-Reckoner*, scorned Aristarchus of Samos for his mathematical misdemeanour in giving a point magnitude could not possibly have tolerated Plato's pseudo-mathematical arguments.[7.51]

Plato had confined practical geometry to military matters, a very limited role in the logistics of warfare, very far from conceiving its use in creating Archimedes' weapons. He had demanded that mathematics be studied for the sake of knowledge itself.[7.52] In contrast, Plutarch did not condemn the use of mechanics for military purposes; he only commented that it lay outside philosophy.[7.53] This difference, together with his comments on the 'celebrated art of mechanics', supports Georgiadou's finding that Plutarch did not always understand Plato.[7.54]

Unless there is supporting evidence elsewhere, Plutarch's view that Archimedes despised his own inventions is an assertion that is dubious in the extreme.

Plutarch's Motives

Drachmann offered an excuse for Plutarch that is quite at variance with that by any other. He started from the proposition that Archimedes tried to express his ideas in mathematical language, but where those ideas were about mechanics and physics they could not always be so expressed, and Plutarch had failed to understand this difference. Drachmann implied that the error was a natural one because no one was to write physics that way for another thousand years.[7.55] This does explain away Plutarch's reference to 'the celebrated and highly prized art of mechanics'. Plutarch could have been merely reporting Plato's views on Eudoxus and Archytas, not necessarily agreeing with them. None the less, it does not explain away enough. There were other writers in the Archimedean tradition who discussed physics, e.g. Ptolemy and Heron. The period of 1000 years is too loose; if Drachmann intended Ibn-al-Haitham, he did not say so, and, if he meant Grosseteste or Roger Bacon, he picked the wrong length of time. As an explanation, it is totally irrelevant to Plutarch's report that Archimedes had to be persuaded to act as an engineer or that he only carried out those tasks for his own amusement or at Hieron's persuasion.

Very recent critics have strengthened the thesis that Plutarch, being a Platonist, was trying to present Archimedes as of that same school. Georgiadou argued that Plutarch's aim in *Marcellus* was to contrast the warlike Marcellus with Archimedes' wish for a life free from war. In order to 'purify' Archimedes for his apparent deviation from the 'proper use of geometry' formulated by Plato, as Plutarch understood those doctrines, Plutarch devised the excuse for the military activities that Archimedes had succumbed to Hieron's persuasion.[7.56]

Phyllis Cullham's analysis complemented Georgiadou's case and came to similar conclusions.[7.57] The contents of *Marcellus 14* had little to do either with the siege of Syracuse or with Marcellus. That paragraph better fitted a philosophical discussion, with the aim of showing that Archimedes wanted to be remembered for his work in pure geometry, emphasizing the symbol on the tomb, that his views matched those of Plato. Since *The Method* directly contradicted this, she had to assume that Plutarch was unaware of it.[7.58] Most importantly, two Platonic ideas in Marcellus had not been properly understood or even recognized before. Whereas Polybius had contrasted the one *psyche* (of Archimedes) with the numbers and resources on the Roman side, Plutarch contrasted that *psyche* with the *soma* of the people of Syracuse, an overtly Platonic contrast between soul and body. When Plutarch called Archimedes *demiourgos*, he did so in a more highly charged way than Polybius had done.[7.59] He was contrasting *doxa*, the belief in the world of appearance, with *episteme*, true knowledge. Hence to admire the Roman task force and Archimedes as a military engineer was, for Plutarch, to succumb to false *doxa*.[7.60]

Within *Marcellus*, Cullham saw a highly ideological manifesto on the proper role and status of the scientist-philosopher in society.[7.61] Archimedes was not an intruder into the biography, he was the man of thought, as against Marcellus, the man of action.[7.62] 'Plutarch's digressions on Archimedes' accomplishments in geometry *betray* [italics added; this is an unusual word in a paper] the real reason for his lack of accuracy about the counter-siege weaponry at Syracuse.'[7.63] Not only was he not interested in military technology, he had a contempt for it. Unfortunately, this led to his simultaneously attributing Syracusan success to Archimedes alone, while insisting that his military devices were without scientific merit.[7.64] In praising Archimedes for his disregard for the active life and excusing his civil and military inventions as due to the persuasion of Hieron, Plutarch was anticipating the neo-Platonism of the next century, and not expressing views that matched Plato's own, particularly in visiting Syracuse for the explicit purpose of educating the then tyrant.[7.65]

Not all Culham's arguments are wholly persuasive. For example, she overemphasized the accuracy of the reporting of the weapon technology by Polybius and Livy. Polybius was as emphatic as Plutarch about Archimedes' crucial role in the siege.[7.66] She allowed nothing for the generally accurate accounts of Archimedes' other activities, especially Plutarch's conveying of the excitement of technology ('the celebrated art of mechanics'), with Archimedes taking his part with Hieron and the Syracusans or Plutarch's failure to condemn the use of mechanics for military purposes.[7.67] There are minor technical deficiencies in her account, too: her comment that Archimedes never used his devices disregards the story of his moving the merchantman.[7.68] Pedantically, *The Method* does not *deal* with mechanics, it *uses* mechanical methods. The third version of how Archimedes met his death is unexplained. Her belief that the consensus of modern scholars rejects Plutarch's version of Archimedes is unwarranted.[7.69] The central theme that Plutarch was using the lives of Archimedes and Marcellus to contrast two very different ways of life is not established. An essay especially written for that purpose would surely have set out the contrast between the two men more fully and made that theme more explicit.

None the less, the accusation by Georgiadou and Cullham that Plutarch was deliberately constructing a Platonic Archimedes has to be taken most seriously, not least because, unusually, they took in the whole essay. Their claim reinforced earlier doubts and explained some of the confusion in Plutarch's account. A defective grasp of Platonism allows for his praise of Archimedes holding to a private life as well as the ambiguity of his comments on using mechanical means to solve geometrical problems. If Plutarch did not know *The Method*, he may not have been aware that Archimedes was at one with Eudoxus, Menaechmus and Archytas. Neither author explained the different emphasis in two sections, the one expressing hostility to engineering and the earlier passage describing his engineering activities as undertaken for amusement. Were Plutarch truly writing as a neo-Platonist, he was a muddled one. None the less, the credibility of Plutarch's record of Archimedes' attitude is made all the more dubious by their criticisms.

Opinions of Other Classical Commentators

Archimedes and the Greeks

Plutarch's opinion needs to be compared with those of many other authors. Hieron, according to the late author Proclus, when Archimedes was able to contrive a mechanical device which enabled a single person to move a large ship, declared that from that day forth everything that Archimedes said was to be believed.[7.70] Proclus added that the same remark was attributed to Gelon, Hieron's son, when Archimedes had solved the problem of the crown. Both anecdotes confirm that these two tyrants of Syracuse were favourable to technology. Indeed, Plutarch makes it abundantly clear that Hieron was always encouraging Archimedes to practical ends, military and civil.[7.71]

Diodorus the Sicilian, nearest Archimedes' time after Polybius, followed the attribution of the invention of the water-screw with a fulsome encomium of Archimedes and his inventions, though Diodorus' 'detailed and precise account' has been lost:

> And a man may well marvel at the inventiveness of the craftsman, in connection not only with this invention but with many other greater ones as well, the fame of which has encompassed the entire inhabited world and of which we shall give a detailed and precise account when we come to the period of Archimedes.[7.72]

Diodorus' attitude is constructive and he gave no hint whatsoever of Archimedes displaying any hostility to invention.

Other writers refer to Archimedes as a mechanician or a practitioner in the mechanical arts.[7.73] The earliest is Athenaios' description, derived from Moschion, probably a contemporary of Hieron, of Archimedes' part in the launching of Hieron's Great Ship.[7.74] The next is by Pappus, who recorded that the ancients called those who studied the mechanical arts *mechanicians*; among that group he numbered Archimedes.[7.75] Both Pappus and Proclus, the latter drawing heavily upon Geminus (first century BC), although they favoured the theoretical side, accepted that 'mathematics' could serve practical ends too.[7.76]

Archimedes and the Romans

Rome recognized his part in holding off Consul Marcellus at the siege of Syracuse for two years. They knew something of his technical achievements. One of his two planetaria, looted by Marcellus, was put on show in Rome; they were described twice by Cicero and praised by the poets Ovid and Claudian and the theologian Lactantius. Cicero, when faced with a seemingly intractable problem, complained that he had a *problema Archimedea,* a problem for Archimedes.[7.77] He excoriated the idea of geometry for its own sake, and regarded it as primarily for practical use.[7.78]

One of the longest and most enthusiastic tributes was that by Silius Italicus.[7.79] After a reference to Archimedes' role in the siege, he went on:

> There was living then in Syracuse a man who sheds immortal glory on his city, a man whose genius far surpassed that of the other sons of earth. He was poor in this world's goods, but to him the secrets of heaven and earth were revealed. He knew the rising sun portended rain when its rays were dull and gloomy; he knew whether the earth is fixed where it hangs in space or shifts its position; he knew the unalterable law by which Ocean surrounds the world with the girdle of its waters; he understood the contest between the moon and the tides, and the ordinance that governs the flow of Father Ocean. Not without reason men believed that Archimedes had counted the sands of this great globe; they said too that he had moved ships and carried great buildings of stone, though drawn by women only, up to a height.

Livy recognized Archimedes not only as an unrivalled observer of the heavens and the stars, but also as someone who was 'more remarkable, however, as inventor and contriver of artillery and engines of war.'[7.80] Pliny also praised Archimedes' contribution to geometry and mechanics.[7.81]

The use of the term mechanic continued into the Empire and beyond. Firmicus Maternus referred to his fellow citizen as among those applying mechanical arts.[7.82] The expression is also found in a letter of Cassiodorus. This use of the term *mechanician* over so many centuries must mean that his commentators, both Greek and Roman, accepted Archimedes as someone who worked in this area, without any sense of its lowering his standing and without any reference to Plutarch's views. In that same letter, Cassiodorus also instructed Boethius to design a water-clock and a sundial.[7.83] Edward Gibbon alone recognized that this meant that Boethius was a practical man: 'he alone was esteemed capable of describing the wonders of a sundial, a water-clock, or a sphere which represented the motions of the planets.'[7.84] It does not matter whether Boethius possessed the practical technical competence to build the water-clock.[7.85] Boethius accepted the call to design the water-clock; Cassiodorus did not think that this was inappropriate work for a Roman Senator.

Archimedes and the Byzantines

Tzetzes, living a thousand years after Plutarch, several times referred to Archimedes and the technical works of him and his successors.[7.86] Tzetzes' own attitude to technology is clear, matched by a complete lack of interest in pure geometry comparable with that of Cicero:

> Geometry is useful for many mechanical works, for lifting of weights, putting ships to sea, rock throwing, and other siege machines, and for setting things on fire by means of mirrors, and other contrivances for defending cities, useful for bridges and harbour-making, and machines that make a wonder in life, such as bronze and wooden and iron things and the rest, drinking, moving, and crying out, and measuring by machines the stades (about 140 km?) of the sea, and the earth by hodometers, and a myriad other works are born of geometry, the all-wise art.[7.87]

The list is remarkably similar to one that might be compiled of those inventions attributed to Archimedes. Most strikingly of all, Tzetzes uttered a variant of Cicero's cry: 'I have need of the machines of Archimedes.'[7.88] There is nothing to suggest that Tzetzes believed Plutarch's account to be a true one.[7.89]

The Attitudes of Other Scientists and Technologists

Plutarch admitted that Eudoxus and Archytas had originated mechanics.[7.90] Contemporaries, admirers and followers of Archimedes did not hesitate to write upon technical matters, notably Ktesibios, Philon of Byzantium and Heron.[7.91] Eratosthenes, Archimedes' correspondent, had a strong interest in the application of geometry to practical matters, an anti-Platonic stance, as Knorr emphasized; Knorr suggested that Archimedes had devised a solution to one problem as a contribution to ship-building.[7.92] Archimedes cannot be taken to be out of sympathy with them.

Vitruvius, in a work devoted to reviewing the engineering practice of his day and explaining its importance to Augustus, compiled a formidable list of those, including Archimedes, who had written commentaries on the subject.[7.93] When Seneca was highly critical of Poseidonius for attributing inventions to intellectuals, he did not enlist in support Archimedes, an obvious candidate if what Plutarch had reported had been true.[7.94] The anonymous author of a work on hydrostatistics, at the end of the fifth century, based upon Archimedes' Principle and referring to him, was still part of a tradition that valued technology.[7.95] Boethius' technology, derived from that of Archimedes, was used to impress the barbarians.

None of the numerous commentators on Archimedes in Classical times referred to or repeated, let alone confirmed, Plutarch's opinion. Vitruvius, and the less reliable Tzetzes, directly contradict his assertion that Archimedes refused to write about his technical discoveries. Plutarch is alone.

Commentary

Plutarch's report of Archimedes' achievements is generally accurate. One comment, that Archimedes took no interest in public affairs, explains his absence from the list of those who were to govern the city during Hieronymus' minority, though it is out of keeping with the achievements of the young inventor of the screw.

The rest is far from convincing. The assertion that Archimedes refused to write technical papers is contradicted by other authors. Vitruvius, a much earlier source than Plutarch, specifically stated he had utilized Archimedes' works. His inventions for military purposes could not sensibly be disclosed before the siege and were impossible to describe after it. Not all his devices would be written out anyway, being more readily demonstrated.

Plutarch's implication that Archimedes was of a like mind to Plato, in objecting to the use of mechanical means to solve geometrical problems, is manifestly false. In *The Method* Archimedes much improved the technique and recommended it to others. He also praised Democritus, a notable opponent of Plato. Macrobius' report of Archimedes' determination of the distances of the planets suggests that he disagreed with the central thesis of Plato's cosmogony. Its mathematical looseness must have been shocking to someone who pilloried a minor mathematical misdemeanour by Aristarchus of Samos in the *Sand-Reckoner*.

If Stillman-Drake accused Plutarch of being a snob and Clagett almost accused him of being a liar, others have been even crueller. They have charged Plutarch with being a Platonist propagandist and an incompetent one at that. In *Marcellus*, they claim, Plutarch aimed to make a critical comparison between Marcellus, the man of action, and Archimedes, the man of thought who wished for a life free from war. For this to be correct, Archimedes had to despise his inventions, but here Plutarch expressed a drift from Plato's own views, exemplified by Plato's attempts to live an active political life, towards the neo-Platonism of the century that followed. Their charge, that Plutarch was constructing what he thought was a Platonic Archimedes without wholly understanding Plato's doctrines, does not take full cognizance of the accuracy of much of Plutarch's reporting on Archimedes and probably underestimates the confusion in his account. None the less, it does illuminate some of the oddities in Plutarch's account: why, unlike Plato, he did not condemn the use of mechanics for military purposes, but simply noted that it lay outside philosophy; the ambiguity of his comments on using mechanical means to solve geometrical problems; his third version of Archimedes' killing;

and his puzzling reference to 'the celebrated and highly prized art of mechanics'.

It also explains why he alone painted him as a 'pure scientist' in contrast to the comments of a goodly number of commentators from Diodorus to Tzetzes. Some, notably Athenaios and Cassiodorus, even refer to him as a *mechanician*, a concept quite contrary to Plutarch's. More particularly, Archimedes' followers and admirers of the Alexandrian school, notably his correspondent Eratosthenes and, much later, Heron, were concerned with applications.

Plutarch's report that Archimedes despised practice does not therefore bear close scrutiny. While his version cannot be wholly disproved, many of his comments are demonstrably incorrect and others are confused. The real Archimedes does not resemble Plutarch's Platonic creation.

The External Evidence: A Hostile *Ethos*?

The other line of defence for Plutarch is to claim that his reported views of Archimedes matched the *ethos* of the Classical period, which was at best indifferent, and at times actively hostile, to technology. Whereas the early Greeks had welcomed technological advances, exemplified in the myth of Prometheus, from the time of Plato and Aristotle onwards, the leaders of society developed an ingrained contempt for those in mechanical occupations, were prejudiced against the making of money, were only active in seeking to maintain things as they were and so were fundamentally opposed to technological progress. As a result, later Classical Antiquity was a period of relative technological stagnation, with few inventions compared with earlier and later periods, and it failed to press ahead with many innovations that were within its grasp. Advances in military technology are an agreed exception, to which ought to be added naval warfare.[7.96]

Two anecdotes demonstrate that hostility to innovation was present. In the first, someone presented his invention of unbreakable glass to the Emperor Tiberius, and when he admitted that he alone had the secret, he was beheaded. In the second, another inventor came to the Emperor Vespasian with a device for transporting heavy columns to the Capitol at small cost. The Emperor rewarded him well but refused to use the invention, saying 'How will it be possible for me to feed the populace?'[7.97] Both inventions could be seen as a threat to society or could have been devised to show that inventions could pose a threat to society. Seneca was highly critical of Poseidonius for attributing inventions to intellectuals, but he may be taken to be criticizing intellectuals, not inventors.[7.98]

Several factors are identified within that hostile *ethos*. No one made systematic efforts to improve machinery or increase production by transferring the consummate skills of Roman public works engineers to the private sector. It was not simply the number of inventions *per se*; there was no general demand for increased productivity or increased production, though Finley acknowledged that the quantitative evidence to determine the complete answer was lacking.[7.99] Even Vitruvius, although devoted to best practice, was concerned with what was feasible, not what was economic. Cato, Varro and Columella in their works on agricultural practice had ignored the need to increase output.[7.100] There was little advance in agriculture, though there was some in arboriculture, for example, the olive tree and grape cultivation.[7.101]Much of the huge expenditure on improvements to the infrastructure, roads, public buildings, water supply, drainage and other amenities was highly wasteful and not aimed at increasing productivity.[7.102] This was particularly unfortunate because venture capital rarely found its way towards creating factories for the handful of large craft industries.[7.103] The first Ptolemies were exceptions; they had improved the revenues of their state and had supported military technology, not least because they feared an idle labour force.[7.104] However, this had been achieved by applying existing knowledge; only water pots for irrigation and the screw-pump were new innovations and their use was severely restricted.[7.105] Advances were at a low technological level.[7.106]

One of the principal causes identified was that theoretical science tended to be an end in itself, with application to mechanical arts thought to be inferior. As a result, there were too few productive links between scientist and technologist and between technologist and craftsman.[7.107] Green was puzzled enough to ask:

> Why was there such a wide gap between theoretical and applied science: the brilliance of intellectual achievement in areas such as pure mathematics, the limitations and poverty of technological development. Why should brains capable of conceiving a heliocentric universe, or of doing pioneer work on conic sections, so signally fail to tackle even the most elementary problems of productivity.[7.108]

He found a failure of technological nerve.[7.109] Others argue that the ancients' grasp of scientific method was inadequate. No one was responsible for science or its application.

How far is the central claim, that Archimedes' views were those of the ruling class, in a society dependent upon slaves and hostile to technology, justified? Was there a relative or absolute failure to invent and innovate? Are the supporting reasons adduced – the absence of venture capital, craft-based industries too small to innovate – accurate and sufficient? Are the ideas of economic push and the desirability of increased productivity absent and was an inadequate scientific method divorced from practice?

Poverty of Invention?

Finley and Green produced very short lists that aimed to show that invention was rare.[7.110] There are other lists, but they are all unsatisfactory; they do not properly distinguish between those inherited, those improved and those originating in the period.[7.111] They omit elementary devices, the wedge, the lever. They leave out many original devices that are hidden within many large-scale developments, such as the ratchet and pawl in catapults.[7.112] Some types of invention are not noted at all; for example, the Julian Calendar that combined administrative with scientific innovation, using the knowledge of Alexandrian astronomers.

White recognized that compiling lists of inventions invited confusion rather than enlightenment. Doing so concentrated too much upon new inventions and ignored the valuable development that took place within existing frameworks.[7.113] An additional objection was that lumping different types of invention together obscured the actual type of advance: the invention of a new tool should be categorized differently from an improvement to an old one and both from a machine tool. An improvement may be as important as a new invention. What matters is the scale of the application: minor developments to an innovation may be more profitable than the innovation itself, an opinion endorsed by a modern capitalist in the form of the Chairman of ICI.[7.114]

To overcome these difficulties, White tackled the exposition of technological advance differently. He reviewed the main areas of innovation and development, then examined the technologies practised within seven substantial activities in some detail: agriculture and food; building; civil engineering and transport; mining and metallurgy; land transport; ships and water transport; and hydraulic engineering. He set out advances in these and some other areas in tables and appendices.[7.115]

In agriculture, his review ranged from hand implements, like the balanced sickle, to the mechanical grain harvester; in food-processing to the water-powered grain mills, olive presses and olive mills; and in water-raising from the inverted siphon, the Archimedean screw, the bucket-wheel and the bucket-chain to the force-pump. In mining, he noted the high profit margins in the Laurion mines of Greece and the elaborate and costly installations in Spain, particular examples being the sets of screws and bucket-wheels in series, first operated by slaves and later by animal power. Metallurgy advanced less, but the Romans added brass and pewter to the alloys already produced; they could separate lead from silver, so making it profitable to re-open the Laurion mines; they improved the refining of gold. The iron tools of Roman carpenters were of high quality. In pottery, the Greeks invented the potter's wheel, stamped Samian ware and the famous black glaze, and the Romans added different colours and glazings. In textiles, White identified improvements to scouring, other cleansing processes, dyeing and finishing. In glass-making and forming, the invention of glass-blowing created a wide range of new glass products. True mosaic, cut cubes in concrete, was a Roman invention and so was its application to walls and ceilings. Underfloor heating was installed in houses and baths. Although the amount of fuel consumed was prodigious, there were identifiable attempts to conserve the heat.

In building and civil engineering, there were innovations in the transport and handling of materials, including a new type of crane; there were improvements in dressing and finishing; there were very many advances in building technology, especially those based upon *pozzalana* hydraulic cement, which was both waterproof and fireproof, and made the joints waterproof too. The kiln-face brick also appeared. The Romans used the new concrete to produce a range of buildings and other structures, extending and developing the vault and the arch in a variety of forms, including, in the Pantheon, a hemispherical concrete dome. Roman civil engineers were able to construct all-weather roads, often over hostile territory, such as marshlands and depressions. They built bridges and viaducts, with different spans; that over the Moselle at Trier survived into the Middle Ages. They mastered the arts of constructing cuttings, embankments, tunnels and aqueducts in very varied conditions. They were able to build dams for a variety of purposes across the Empire. They made few advances in vehicles, but their transport system was unmatched for millennia to come. Among the numerous innovations at sea were the anchor, the lateen sail, harbour works, including artificial ports, lighthouses with mirrors and dry docks. In hydraulic engineering, an efficient system of water supplies was established and new devices, notably the hydraulic valve, were developed.[7.116]

White was not able to be comprehensive in his coverage. He dwelt very little on the surveying skills of the Roman engineers. Parchment was omitted; its introduction was to be of critical importance to the survival of learning in both the West, after the fall of the Empire, and Byzantium.[7.117] Another interesting omission was the threaded nut and bolt.[7.118] Rome developed and tried to enforce building regulations.[7.119] The decline in size of farm animals after the collapse of the West is at least negative evidence that the Romans paid attention to breeding.[7.120]

White's survey of Greek and Roman technology demonstrated that Classical Antiquity made a large number of inventions of its own, improved many that it received from its predecessors and was ready also to adopt those made by the barbarians. It showed that the period up to Archimedes' death was far from barren of innovation and invention and the range of those applications over the whole period was substantial.[7.121] Indeed, Oleson concluded that the substantial advances in three areas, water-lifting devices, warfare and building, warranted the term 'revolution'.[7.122] Ward-Perkins added architecture and Hill added civil engineering.[7.123]

An innovative spirit survived until the end in the West: the onager was an attempt, by a technologically aware army, to compensate for its inability to construct the more powerful catapults.[7.124] Long after the imposition of the Diocletian strait-jacket, the water-driven mill spread during the late Western Empire, and almost at its end an inventor was able to devise what was possibly the most extraordinary invention of the Classical world, a water-driven saw capable of cutting marble; what is more, administrators had enough flexibility to put it to work.[7.125] The Eastern Roman Empire immediately adopted Kallinokos' (seventh century AD) invention of Greek fire, not least because the situation of Byzantium, under siege from the Arabs, was desperate.[7.126]

Wikander, in endorsing White's conclusions, drew attention to the increasing amount of evidence that the Romans had high standards of technological application over a wide range of activities, and that that range has grown with the increased effort put into investigating their contributions. Much of the apparent remaining barrenness may be due to the bias in Classical literature.[7.127] Archaeology had so far played a restricted part in correcting the picture because technological artefacts were difficult to identify.[7.128] Additional justification had already appeared in the discovery of a Byzantine gearing system.[7.129]

Innovations

Athens, in its heyday from the sixth to the fourth centuries, produced many innovations funded in part by its technical ability to exploit its silver mines. Its exports of pottery and olive oil were examples of large-scale trade that stimulated the revolution in navigation and the shipping trade.[7.130] Its public building programme led to some advances in building technology.[7.131] Its theatre was unsurpassed and theatrical machinery advanced too, funded by the state, largely through encouraging its wealthiest citizens.[7.132] When her power faded, the number of innovations declined.

A new impetus appeared with the rich centralized states and the importance of material progress by certain of their rulers. In Syracuse, Hieron could allow Archimedes his head. At the same period in Egypt, under the first, magnificent, Ptolemies, there were two major areas of advance, warfare and water-lifting. The encouragement and support given to philosophers and scientists, particularly at Alexandria, Syracuse and Pergamon, contributed greatly to technological development in the Greek world.[7.133] Under the *Pax Romana* innovation found even fuller expression.

When the Romans conquered the Greek East and drew upon its wealth, they began to refashion Rome and to provide the infrastructure of the Empire. The building programme devised by Agrippa, and encouraged by Augustus, coincided with not merely a substantial number of innovations, but also their application in construction; they produced revolutions in building technology and architecture.[7.134] These changes, which extended over two centuries, would be very difficult to understand unless there were people in authority ready to adopt them. It is wholly unreasonable to suppose that a hostile *ethos* hindered Agrippa, with the backing of Augustus, from adopting innovations to push through that large-scale state-supported building programme.[7.135] Such a programme could very well have led to a shortage of skilled labour. This would have stimulated innovations that, though labour-intensive, would provide employment for the unskilled, dispossessed peasants who had migrated to Rome.[7.136]

Advances in the Roman period, as with a large proportion of those in the earlier period, were almost all linked with the state. These included the introduction of window glass in Roman army camps and new ways of building roads, bridges and sewers, so creating and improving the infrastructure of the Roman state. The state had to ensure supplies of grain. The organization that lay behind the transport of corn from Egypt to Rome – the ships, the harbours, the docks and the lighthouses – was prompted by the demands of large-scale trade that produced large-scale innovations to match. New styles of pottery were introduced, blown by warmed air, while statuettes were produced on assembly lines.[7.137] The large-scale production of pots at low prices, when an average of 30,000 objects could be fired at one time in one kiln, could be termed industrial production.[7.138] Of two late innovations, the Barbegal flour mills demonstrated that the mechanized technology of large-scale operations was within the capacity of the Roman state,[7.139] and the water-driven saw-mill for cutting marble at the imperial capital of Trier was also likely to have been developed by the state officials.

Other innovations spread within the security of the state, e.g. the comparatively rapid spread of the products of the discovery of glass-blowing led to a plentiful supply of glass vessels.[7.140] The spread of the water-mill and adoption of the mechanical harvester in Northern Gaul were the work of the great landowners, and the Emperor was a great landowner, who recognized their advantages. Oleson considered that three technologies – military, water-lifting and building – were crucial to the maintenance of the state. To them, from the Principate onwards, ought to be added mining and shipping. Here, too, the Romans introduced many inventions and added substantially to technology. Demand could be met not only without changing the basic organization of, but also by supporting, that society.[7.141] Economic forces influenced the shift of production of pottery and food-growing from country to country. The invention of parchment augmented the supply of papyrus for a society based upon literacy; the reform of the calendar was required by the needs of a fundamentally agricultural society. Other changes, such as the development of larger-scale production units in cities, as opposed to those on the big estates or those under the control of the army, would have demanded social changes that the ruling class might have suppressed as soon as they realized what was happening.

The delays in applying some of these innovations have occasioned too much surprise; it was not necessarily owing to a hostle *ethos*. Commentators have not only ignored, but also seem to be ignorant of, the nature of engineering development. The problems in converting Archimedes' design of a water-screw to a working piece of equipment went unrecognized; they are the basis of Drachmann's realization of the difference between the over-elaborate device described by Vitruvius and the much simpler working versions, with their design varied to match the use to which they are being put. The water-mill spread far more widely than has been credited. If the time taken for its general use to become widespread appears to be long, this was partly because of the need to overcome defects in the original design and to learn how to operate them in many different conditions: different streams with different water flows, different rates of sedimentation and different climates giving rise to variable flows. To transfer a design of a water-mill from one area to another demands a good understanding of the factors that determine its efficiency and whether it is worth installing. Yet these elementary technical and economic issues went unremarked until noticed by Wikander.[7.142] A comparable example is the wheelbarrow. It was probably invented in Classical Greece, it survived the fall of Rome only in the Byzantine Empire and it did not reappear in the West until the late twelfth century. Yet, though it is labour-saving and versatile, its use was limited to England, France and the Low Countries. Only in the fifteenth century did it spread more widely; here the reason for the delay may have been social inertia.[7.143]

Denouncing the ancients' failure to turn Heron's steam toys into a working steam engine is perhaps the most extreme and superficial form of the belief that classical antiquity neglected the opportunity of developing an independent prime mover.[7.144] Needham's review of the pre-history of the steam engine may be over-elaborate in its identification of all the influences that were required to produce a working engine, but the mere fact that two hundred years elapsed between the recovery of Heron's work and Savery's practical, but grossly inefficient, engine should give some pause for thought.[7.145] And the fact that the first steam turbine, as opposed to a steam engine, did not appear for another two hundred years illustrates that the appearance of an invention does not depend solely upon the components being available. Someone must have the idea and the determination to put that idea into practice and find the capital and the market. Watt found Boulton. Both had to find Wilkinson.

Finley did not consider sufficiently the effect upon productivity of those inventions that he did accept, although this factor is difficult to measure even in modern times, when it is considerably higher.[7.146] His belief that the many improvements to the infrastructure, such as expenditure on roads and on water supplies, had no effect on productivity is highly improbable. He ignored their role in the maintenance of the system of government, which was just as necessary. Possibly, it contributed to the survival of the Roman Empire for longer than might have been expected, given the burden of its defence. Although Oleson did not attempt to quantify the effects of the introduction of new water-lifting devices, these irrigation techniques helped to increase the production of corn and made Egypt the source of supply for Rome. Certainly, that supply, part of the policy of 'bread and circuses', was essential for any emperor's survival and produced a thoroughgoing revolution in the supply system, from harbours to ships.[7.147]

This evidence about the amount of innovation does not wholly refute Finley's specific thrust about the absence of technological push where it might be expected to be found in Vitruvius, and his more doubtful assertion about its absence from agricultural handbooks, but Frontinus devoted a work to the management of water supplies and improvements were made in animal and crop husbandry.[7.148] Some of the craft industries were large enough to warrant expenditure on innovations, notably the pottery industry. Finley ignored or underestimated the role of the state in producing venture capital and applying innovations. There may be little evidence of a general technological push, but there is ample evidence of particular ones. The intensity of application was, as it has remained, uneven.[7.149] As a generalization, the Roman state and its enterprises provided the main thrust for the likelihood of an innovation being applied, just as previously the rich Athenian democracy and the powerful Hellenistic tyrants and kings had done. The assumption of a general hostility or indifference to technical advance and the absence of technical enterprise in Classical Antiquity is untenable.

Finally, the main underlying reason, that an inadequate scientific method divorced from technological and technical practice handicapped invention, is not sustainable. To modify De Solla Price's metaphor, the belief that science, technology and the inventor largely danced separately, or, more prosaically, were not sufficiently advanced to help each other, relies on an unconscious acceptance of a simple model of innovation that assumes a linear transfer from science through various stages to increased productivity.[7.150] The list of inventions inherited from previous civilizations and those attributed to the Celts shows that this model was wholly inappropriate; in this period, as they would be for many hundreds of years to come, invention and innovation were almost entirely independent of science.

There were exceptions. Military innovations drew heavily on the efforts of locally supported research workers in applied science and mechanics working with craftsmen. Dionysius of Syracuse combed the Hellenistic world for skilled craftsmen and inventors, paying them handsomely in money and attention and receiving his reward with the invention of the *gastraphetes*.[7.151] The early Ptolemies also reaped the rewards of their support of the research workers at the Museum of Alexandria and the links they established between the innovators and application, notably the formulae for setting the ranges of *ballistae* and the invention of water-lifting devices for more effective irrigation, two areas where Archimedes made a powerful contribution. The Julian Calendar was another successful combination of scientific knowledge and administrative necessity.[7.152]

Skilled craftsmen were active in 'high' technology. They made the sphere or planetarium, which Archimedes had designed, the Antikythera mechanism and the Byzantine calendar.[7.153] It is impossible to believe that Archimedes did not have access to skilled craftsmen at Syracuse when he carried out his feats; in fact, Athenaios' account of the Great Ship reveals that he could call upon them. The armies, the shipyards and the mines must have trained their own craftsmen, as did Crassus' corps of fire-fighters in Rome; training on the job remained the rule until recent times. For specialist, professional training students had to go, as Archimedes had gone, to Alexandria. Training facilities for architects and civil engineers must have existed, because when Constantine's plans for building Constantinople were held up by a shortage, their hurried rehabilitation was decreed.[7.154] Even were the general grasp of scientific method uncertain, the understanding of basic principles would still be a great help in engineering practice; for example, Archimedes' ability to calculate mechanical advantage.[7.155]

The Ethos

The most puzzling thing about the attempts to prove that the Classical period was hostile or at best indifferent to technological change is that it is refuted by Plutarch himself. Hieron actively encouraged Archimedes in his engineering activities and did so partly to educate the Syracusans.[7.156] The stories about the Great Ship and the other public activities of Archimedes are as much Hieron as Archimedes at play. Defenders of Plutarch ought to be arguing that Archimedes was out of sympathy with that particular *ethos* of Syracuse. Instead, those who appear unaware of Hieron's attitude argue that Plutarch expressed the prevailing *ethos* of the Classical period. Plato and his followers were hostile to the mechanical arts and to technological advance, and the other schools were indifferent or ineffective. Anecdotes about the Emperors Tiberius and Vespasian having rejected innovations also exemplify hostility to changes, especially those likely to produce changes to the structure of society.

This does not justify the belief that Archimedes held similar opinions. The opposite attitude existed, as must be obvious from the commentaries on Archimedes. Opinions hostile to technology expressed by the literary classes were not shared by the population as a whole.[7.157] Legislative texts, particularly from Athens, show that manual dexterity and skilled labour were held in high regard.[7.158] It is not by chance that the change in literary attitudes coincided with the change in Athens' circumstances. As its power and independence faded, the number and range of its innovations declined.

The Elder Pliny complained that much too little attention was being paid to invention in his day, though he lived at the time when the Romans were revolutionizing building technology and were shortly to revolutionize architecture.[7.159] Vitruvius appeared confident that the Emperor Augustus was interested in his book. The revolution in building technology was directly encouraged by Augustus' programme for rebuilding Rome.[7.160] The Emperor Hadrian was trained as an engineer and the part engineering played in the maintenance of the strength of the Roman army for so long is proof that practical men played an important role in society. The names of the architects and engineers that have survived, often with descriptions of their achievements, is negative evidence that the importance of their technologies was recognized.[7.161] Best known today are Anthemius of Tralles, a mathematician and commentator upon Archimedes, and Isodore of Miletus, the architects of Santa Sophia. Justinian's contributions to architectural and engineering problems, attributed to him by Procopius, are only credible as flattery, but they show that the ability to solve such problems was regarded highly.[7.162]

Some have argued that the Alexandrian tradition was different.[7.163] It was cosmopolitan, part Egyptian, part Greek, with a liberal sprinkling of other groups, such as the scholars and traders who came from as far as India. Greek was the language, not the race, of the mathematicians and the different traditions produced cross-fertilization.[7.164] However, Tarn had earlier put the generally accepted view that the Greeks maintained themselves as a governing race from which the natives were excluded.[7.165] The argument of an Alexandrian tradition different from that of the Hellenic and the rest of the Hellenistic world is unconvincing, apart from the way it is interpreted; Kline thought Archimedes was part of the Alexandrian tradition, and Hacker and Joseph thought the opposite.

Innovation to match the demands of economic and social forces, then as now, depended upon opportunity and necessity. Where the ruler or ruling class recognized the advantages of innovation, so he or they provided venture capital to inventions and for their application. There was no unified *ethos* over the whole period, the attitude varied from place to place over time.[7.166] Support for innovation moved with power and authority, flourishing at times at Athens, Syracuse, Pergamon, Egypt and then in Rome.

Comparing Achievements

Duval praised the Romans' achievements in the technological domain, where they showed a rare power of accomplishment and remarkable powers of dissemination in spreading techniques, in developing existing inventions and in their general practical, efficient and economic application on large and small scale over their Empire. However, he concluded from his survey that they were weak in the mechanization of industry and in the application of science to the development of machines and to agricultural production. Consequently, they were no better able than their predecessors to bring about an agricultural or an industrial revolution.[7.167] This critique has invited a comparison with other periods, notably the late Dark and early Middle Ages, and has led to the thesis that it was in these periods that the West saw an increase in the application of innovation and inventions and a change in attitude towards encouraging economic growth.[7.168] Oleson, much influenced by Lynn White, concluded that the Classical period had had fewer inventions than the Middle Ages.[7.169]

It is generally accepted that ancient technology made relatively little use of the inanimate power of wind and water. Thus, although they were familiar with wind power through ships' sails, they never used windmills. Again, they never constructed water-wheels in the numbers they might have done.[7.170] Yet, while Wikander was able to demonstrate from the archaeological record that their usage had been very much greater than believed, it still remains puzzling why their use was confined to grain-milling, with only one authentic exception, the marble-cutting saw-mill at Trier.[7.171] Gimpel identified the Cistercian policy of exploiting water power as one of the driving forces behind this change and correctly emphasized that there was no evidence of a directly similar policy in the Roman world.[7.172] Undoubtedly, water-mills, for example, begin to be used for purposes other than grinding grain by the ninth century. The use of the semi-automatic saw-mill for cutting wood had to be restricted because of its effects on timber supplies. On the other hand, there is little evidence about when these inventions and innovations were introduced, as opposed to when they are first recorded. Much of medieval technology was derived from Classical and barbarian predecessors and little was truly innovative for a very long time.[7.173]

This particular criticism of the Classical world for its failure to innovate is a curious phenomenon. The failure of the Chinese civilization to produce the scientific, followed by the industrial and agricultural, revolutions is usually the subject of sympathetic investigation and understanding, while the failure of the Islamic civilization to anticipate any of the three revolutions is hardly thought worthy of comment. If the failure of the Roman world to develop the steam engine is a legitimate question, why, given the optical knowledge of Ibn-al-Haitham and his successors and the skills of their opticians, does no one ask why the Muslim world did not invent spectacles?

It also raises questions of selectivity. The protagonists often assume that because something is not mentioned in the literary record, it cannot have existed; the Antikythera mechanism is only one example of something totally unexpected. Another, almost intolerable, error is to assume that an invention, the heavy plough, took place far later than it did.[7.174] So many Greek and Roman technological developments, as Landels has demonstrated, actually worked: Ktesibios' force-pump was still in use in the nineteenth century. Slavery was expensive, not cheap; it had not inhibited the introduction of labour-saving technology introduced into Pharaonic and Ptolemaic Egypt.[7.175] Yet another fallacy is to assume that the standards of the two periods were equal. Roman quality control was generally effective; notably, in producing cement masonry of high quality, they took good care to ensure that all the air was squeezed out. Later ages lost this professional conscientiousness: medieval cement is distinguished from its predecessor by its lack of solidity, its poor quality and the cavities left in the cement and between the stones owing to poor mixing and insufficient compression.[7.176] Most strikingly of all, the Romans' capacity to manage a range of large-scale technological projects disappeared.[7.177] Moreover, much of their technology was of the kind now named 'appropriate'.[7.178] Given the environmental damage inflicted upon the Mediterranean world by the exploitation of that 'appropriate technology', there is a case for being thankful that their technology advanced no further. Gimpel had no hesitation in concluding that the Middle Ages had suffered from the over-development of some technologies and that the stagnation in the dreadful fourteenth century was in part because of environmental exhaustion.[7.179] Hollister-Short found that the forests of much of Central and Western Europe were devastated by the demand for timber in the centuries to come.[7.180]

In comparisons of civilizations, the quality of life for their citizens is an important factor. The application of technology in the Roman world produced a remarkable infrastructure. A Roman senator in his villa and his town house was much more comfortable than the Norman knight in his castle: he had windows, central heating, good wholesome water supplies and much more comfortable furniture, including a bed with flexible springs; he had private and public baths, paved streets and firefighters.[7.181] In most of those respects, his conditions were better than those of Victorian London. He could move on roads that were not equalled for millennia at a speed that was not matched for just as long a period. If the citizens of Roman towns were always at risk from starvation, then so were the townsmen of the successor states; the last famine in Scotland just preceded the Act of Union of 1707.[7.182]

Much of the argument is therefore misplaced. Before a properly based comparison can be made, agreement has to be reached on the 'outcome measures' that are applicable. The information to produce more definitive answers is lacking, in particular to the central enquiry as to whether the Classical world was less innovative technologically than it could have been, and whether it was less innovative than other periods. Fortunately, more rational answers to both questions may be forthcoming as the work of pioneers is extended.[7.183]

None the less, were invention and innovation the children of opportunity, then that opportunity was often lacking in what was basically a controlled society. There was little scope for the development of an entrepreneurial economy in the Roman world. It was not the existence of slavery, but the fact that, with the establishment of the Principate, and certainly after Diocletian's 'reforms', few were free to act independently. That system collapsed with the disappearance of the Western Roman Empire and the state ceased to be the main way of encouraging, determining and implementing innovation.

Also missing from Classical Antiquity was the almost messianic attitude towards technological progress and its consequences for society, expressed by medieval figures such as Roger Bacon and Guillaume de Ste Cloude. The fact of its existence being accepted, however reluctantly, by Petrarch and the frequency with which these ideas were repeated after them and echoed later by Francis Bacon have only a modest counterpart in the literary record that survives from Classical Antiquity. Commentators on Archimedes the *mechanician* praised his inventiveness; the Elder Pliny welcomed inventions and bemoaned the falling off in their number; Vitruvius praised inventors, but unlike his modern counterparts did not specify areas where more information was required.[7.184] Was the anonymous fourth-century author of *De rebus bellicis* the only person to write about the value of technology, and the only person who made a modest, but conscious, effort to improve the technology of his day?[7.185]

COMMENTARY

Inventions and Achievements

Almost all the civil engineering achievements attributed to Archimedes are heavily dependent upon secondary sources, written centuries after his time. This makes the establishment of their plausibility extremely difficult. None the less, that Archimedes conceived, designed and built planetaria is well authenticated. Not only do a number of references survive, admittedly not always entirely independent of each other, but the discovery and elucidation of the Antikythera mechanism and the

London Sundial Calendar have demonstrated that more elaborate devices of the same kind were constructed by his successors. That Archimedes also designed the water-screw that bears his name is almost as probable. Vitruvius' description may well derive from Archimedes' own. Later versions, modified by practical men for specified purposes, reveal the normal process of engineering development.

The other two achievements examined are both less substantial and less substantiated. The cry 'Give me a place to stand and I will move the earth' may only express Archimedes' confidence in his understanding of the principle of the lever. If, as Plutarch recorded, Archimedes was commanded by Hieron to demonstrate that principle, then strictly Plutarch's account of launching a ship is irrelevant, because it nowhere mentions the use of a lever, although the devices that are mentioned do utilize mechanical advantage effectively. Besides, the two surviving versions are so different that they suggest that neither is more than a romantic reconstruction of something simpler, but still astonishing to the uninitiated. Possibly, Archimedes demonstrated a system giving him a remarkably high mechanical advantage.

Arts of War

The evidence that, at Hieron's insistence, Archimedes prepared the defences of Syracuse well before the Romans besieged the city is fully documented in the authoritative accounts in Polybius, Livy and Plutarch. Statements by Finley and others that the weapons were improvised near or at the time of the siege are incorrect. The weapons, his deployment and use of them, and most particularly their totally unexpected effect on the Roman attackers are described by those three authors and confirmed by references in other writers. To describe them, as some have done, as fabulous or legendary is to abuse the meaning of words.

On the contrary, there is little evidence, as White concluded, that Archimedes had developed new military inventions. He may well have produced improvements and ensured that they were effective for their purpose. Even his devising of catapults having different ranges was based on research carried out at Alexandria by the Ptolemies. What was 'original' was the way in which he used those weapons; what was 'fabulous' was the way in which they were selected so that the attacker, whether approaching by sea or by land, despite his having slipped through one set of weapons was still exposed to those capable of working at shorter ranges. Although Polybius realized this, and deserves due credit for his understanding – that the essence of Archimedes' achievement was to provide artillery that was comprehensive in its depth of coverage – there is no sign that the lesson was understood by the Romans.

Attitude to Engineering

Archimedes' genius as a mathematician, and paradoxically Plutarch is one of the few to recognize it, inspired few followers despite his hopes to the contrary. He was, however, greatly admired in Antiquity for his mechanical inventions and that admiration was part of the legacy that came to the fore quite early in the revival of learning. Late Renaissance engineers, as Gille emphasized, thought of themselves as working within the Archimedean tradition. It was only when Plutarch's works reappeared in the fifteenth century that the idea that Archimedes was a sort of Platonic idealist became current. Once established, it often dominated the minds of many whose exposition, directly or indirectly, demonstrated that there was every reason to be sceptical about its truth.

Plutarch had some technical understanding: his report of Archimedes' mathematical genius and his technical achievements cannot be challenged, while his comment that Archimedes took no interest in public affairs does explain his taking no part in the affairs of Syracuse in the chaos of the post-Hieron period. Otherwise, Plutarch's principal assertion that Archimedes despised engineering practice and refused to write technical papers does not stand. Archimedes did, though he may not have described those that were better demonstrated and he would not have disclosed those that were for military purposes. However, in so far as Plutarch implied that Archimedes followed Plato in objecting to the use of mechanical means to solve geometrical problems, *The Method* reveals that the reverse was true: Archimedes extended and much improved the technique and recommended it to others. Plato's pseudo-mathematical cosmogony must have been intolerable to the man who represents the epitome of mathematical rigour.

The evidence that Plutarch was acting as an incompetent Platonist propagandist in constructing what he thought was a Platonic Archimedes is substantial. Although this charge probably underestimated the confusion in much of his reporting, it illuminates some of the oddities in Plutarch's account and explains why he alone, of all the many commentators on Archimedes, painted him as a 'pure scientist'.

The alternative or complementary defence for Plutarch is to claim that these views were typical of the general hostility to technology of the Classical period. This is immediately refuted by Plutarch himself, who recorded that Hieron encouraged and persuaded Archimedes to apply his abilities to practical affairs. While there was undoubtedly some hostility to change, certainly after Athens had lost its independence of action, some of the rulers of the new centralized states, notably the early Ptolemies, encouraged innovation, while under the Principate there were revolutionary advances in technology in areas fundamental to maintenance of the Roman state.

Plutarch's assertion contains too many discrepancies and contradictions to be maintained as a serious proposition. To invoke him in support of Archimedes' attitude is to assume the truth of what is to be proved; to convert his statement to a generality about Classical times is unwarranted. No substantial reasons have been offered to suggest that Plutarch's opinions are typical of his class in decline, living in a Greece that was a shadow of its former self.

As for Archimedes' own attitude, the inventions attributed to Archimedes are too many, too varied and too brilliantly conceived for

Plutarch's comment to be credible. They and the achievements attributed to him show a man with a practical bent, and his active part in the repelling the first Roman assault suggests he enjoyed the practice of his art. The evidence that Archimedes thought and acted in the same way as his peers and that he was part of an applied scientific or mechanical tradition is substantial. Perhaps he typified part of that tradition too well. Since Syracuse switched sides following the death of Hieron, his patron, he apparently made no distinction between either party in his city-state, like many another chief scientific adviser after him, including his great admirer Leonardo. 'He kept the true faith of an Armourer.'[8.1]

The survey by Knorr of Archimedes the man ended somewhat ruefully with the acknowledgement that Dijksterhuis had made no attempt to offer an opinion and so he too would refrain.[8.2] It would be brave or foolhardy or both to venture beyond these two authorities. Nothing presented here, however, provides any reason to suppose that Archimedes thought his engineering inventions were as important as his mathematical discoveries. Plutarch recorded, and Cicero confirmed, that his tomb had a device showing the ratios of the volumes and surface areas of the sphere and its circumscribing cylinder.[8.3] Plutarch could hardly have invented the story that the symbol had been chosen because Archimedes declared it was his favourite demonstration. But supposing Newton had been asked the same question, would his answer have been the works on alchemy or the *Principia Mathematica*?

Notes and References

1.1. Common English usage does not distinguish between engineer and technologist; nor does it distinguish between the professional engineer and the craftsman. The same term serves for both. It is, however, customary to name the branches of the subject, with the appropriate adjective being military (but naval architecture), civil, structural, aeronautical, etc. There are exceptions: it is building and fire technology, though fire engineering is also used. A distinction sometimes attempted is that a technologist applies scientific knowledge to engineering problems. On that definition, Archimedes was a technologist, but the phrase *Archimedes the Engineer* sounds better.

1.2. A. Suarès, *Hélène chez Archimède* (Paris, 1955) (woodcut illustrations by Picasso).

1.3. T.L. Heath, *The Works of Archimedes* (London, 1897), Dover edn (New York, 1953), xv and 223 (*Sand-Reckoner*).

1.4. Plutarch's *Parallel Lives*, part trans. I. Scott-Kilvert as *Makers of Rome: Marcellus*, 14 (Harmondsworth, 1965), 85–118, 99 (kinsman and friend); cited hereinafter as *Marc., op. cit.* (1.4). Heath, *op. cit.* (1.3), 221.

1.5. R.J.A. Wilson, *Sicily under the Roman Empire: The Archaeology of a Roman Province 36* BC *to 535* AD (Warminster, Wiltshire, 1990), 60.

1.6. Titus Livy, *Ab urbe condita*, part trans. A. de Selincourt as *The War with Hannibal* (Harmondsworth, 1965), 24.4, 235–6; cited hereinafter as Livy, *op. cit.* (1.6).

1.7. Heath, *op. cit.* (1.3), xvi and 232 (*Quadrature of the Parabola*, Conon my friend).

1.8. *Diodorus of Sicily*, 5.37.3, trans. C.H. Oldfather (London, 1961), 3:199.

1.9. H. Rose and S. Rose, *Science and Society* (London, 1969), xii.

1.10. Polybius, *The Histories*, part trans. I. Scott-Kilvert as *The Rise of the Roman Empire* (Harmondsworth, 1979), 8.3–7, 367–8; or *The Histories*, trans. W.R. Paton (London, 1922), 3:461. Normally quotations are from Scott-Kilvert as being the more readable; cited hereinafter as Polybius, *The Rise, op. cit.* (1.10); Livy, *op. cit.* (1.6), 25.31.9–10, 338; 24.34.13, 273; *Marc., op. cit.* (1.4) 14.7–9, 98; Livy, *loc. cit.*, 338.

1.11. Livy, *op. cit.*, (1.6), 338; *Marc., op. cit.* (1.4), 19, 104.

1.12. Polybius, *Rise, op. cit.* (1.10), 8.7, 368; *Diodorus of Sicily*, trans. F.R. Walton (London, 1957), *op. cit.* (1.8), 11: 193 (excerpt from Tzetzes, *Hist.*, 2.35.105). No modern edition of Johannes Tzetzes' works exists; his references to Archimedes are scattered in translations of other authors.

1.13. Cicero, *Against Verres*, 2.4.58, 131–2, trans. H.G. Greenwood (London, 1935), 2: 441; Livy, *op. cit.* (1.6), 25.31, 338; *Marc., op. cit.* (1.4), 19, 104; Pliny, *Natural History*, 7.37, 1.125, trans. H. Rackham (London, 1942), 2: 589–61; Firmicus Maternus, *Mathesos Libri VIII*, trans. Jean Rhys Braun as *Ancient Astrology Theory and Practice* (New Jersey, 1975), 210.

1.14. *Marc., op. cit.* (1.4), 17.7, 105; *Plutarch's Lives: Cicero*, trans. J. Dryden, rev. A.H. Clough (London, 1939), 3: 186–226; D.L. Simms, 'The Trail for Archimedes's Tomb', *J. Warburg and Courtauld Institutes*, 1990, 53: 281–6, 285.

1.15. J.J. Bernoulli, *Graechische Ikonographie*, two volumes (Munich, 1901), 2: 178–9; G.M.A. Richter, *The Portraits of the Greeks*, three volumes (London, 1965, suppl. 1974), 2: 245.

1.16. V. Cronin, *The Golden Honeycomb* (London, 1985), 116. D.L. Simms, 'Archimedes and the Invention of Artillery and Gunpowder', *Technology and Culture*, 1987, 28: 67–79; journal cited hereinafter as *T & C.* D.L. Simms, 'Archimedes' Weapons of War and Leonardo', *BJHS*, 1988, 21: 195–210.

2.1. *Marc., op. cit.* (1.4), 14.7–9, 99.

2.2. G. Sarton, *A History of Science*, two volumes (Cambridge, MA, 1959), 2: 122–5, 122; description from Athenaios, *Deipnosophists, 5.203–209*, ed. C.B. Gulick (Cambridge, 1928), 2: 421–47 (Athenaios claimed to be quoting from Moschion, whom Sarton thought was probably a contemporary of Hieron).

2.3. Silius Italicus, *Punica*, 14, 337–55, trans. J.D. Duff (London, 1934), 2: 297–9.

2.4. Proclus, *Commentary on the First Book of Euclid's Elements*, 3, 63–4, trans. intro. notes J.R. Morrow (Princeton, NJ, 1970), 51.

2.5. Oribasios, *Collectii medicarum reliquiae*, 44.22, ed. J.J. Reader (Leipzig, 1933), 4.

2.6. E.J. Dijksterhuis, *Archimedes*, trans. C. Dikshoorn (Princeton, NJ, 1987). With new bibliographic essay: W.B. Knorr, 'Archimedes after Dijksterhuis: A Guide to Recent Studies', 15–16. Diodorus, *op. cit.* (1.14), 11: 193 (Tzetzes, *Chil.*, 2, 103–49).

2.7. B. Cottrell and J. Kammingana, *Mechanics of Pre-industrial Technology* (Cambridge, 1992), 216.

2.8. Dijksterhuis, *op. cit.* (2.6), 15.

2.9. A.G. Drachmann, 'How Archimedes Expected to Move the Earth', *Centaurus* 1956–8, 5: 278–81. A.G. Drachmann, 'The Crank in Graeco-Roman Antiquity', in *Changing Perspectives in the History of Science*, ed. M. Teich and R. Young (London, 1973), 48–51.

2.10. Drachmann, 1956–8 (2.9), 279 (Drachmann cannot have read Athenaios carefully; he thought that Archimedes supervised the building of the ship), 1973 (2.9), 48.

2.11. Drachmann, 1956–8 (2.9), 279–80; 1973 (2.9), 51 (without this limitation).

2.12. Drachmann, 1956–8 (2.9), 279 (Archimedes invented endless screw). Drachmann took the screw from Athenaios, though the text gives the equivalent to windlass, but asserted that this was corroborated by Plutarch's claim that it was done easily; the assertion is without merit; *loc. cit.*, 280 (compound pulley is attributed to Archimedes by Oribasios, Plutarch and Tzetzes), probably falsely.

2.13. Drachmann, 1973 (2.9), 51.

2.14. *Ibid.*, 42; Heron d'Alexandre, *Les Mechaniques ou elevateur des corps lourds*, trans. from Arabic by B. Carra de Vaux, intro. D.R. Hill, comment A.G. Drachmann (Paris, 1988), 226.

2.15. Drachmann, 1973 (2.9), 51. Proclus, *op. cit.* (2.4), 51. Proclus added that Gelon made the same remark when, without destroying the crown that had been made, Archimedes discovered the weight of its component materials.

2.16. Drachmann's second device has no pulleys, in complete disregard of Plutarch's statement.

2.17. Triremes beached during the night and at various other times to dry out; their crews were capable of pulling them off and on shore. A.M.H. Jones, *The Later Roman Empire*, two volumes (Oxford, 1964), 2: 843–4 (Moschion on beached ships; 3000 men on job). Herodotus, *Hist.*, 6.24. See W.S. Lindsay, *History of Merchant Shipping and Ancient Commerce*, four volumes (London, 1874), 1: 76–7.

2.18. Cotterell and Kamminga, *op. cit.* (2.7), 27 (known to Greeks, must have been known to Egyptians).

2.19. D.R. Hill (private communication).

2.20. A.W. Sleeswyk, 'Archimedes' Odometer and Waterclock', in *Ancient Technology*, Finnish Institute at Athens Symposium, 30 March to 4 April 1987 (Helsinki, 1990), 23–37.

2.21. M.J.T. Lewis, 'The South-pointing Chariot in Rome: Gearing in China and the West', *Hist. Tech.*, 1992, 14: 77–99, 88, 95.

2.22. Sleeswyk, *op. cit.* (2.20), 28. Lewis, *op. cit.* (2.21), 93.

2.23. Jones, *op. cit.* (2.17), 2: 843–4 (largest vessel – reliable record is 330,000 kg). Cotterell and Kamminga, *op. cit.* (2.7), 28 (frictional forces of like materials are similar).

2.24. Cotterell and Kamminga, *ibid.*, 93.

2.25. *Ibid.*, 41.

2.26. F. Welsh, *Building the Trireme* (London, 1988), 154–5, 163.

2.27. Polybius, *Rise, op. cit.* (1.11), 8.5–6, 366–7.

2.28. Sarton, *op. cit.* (2.2), 1: 120.

2.29. *Ibid.*, 2: 363 (port of Rome had *navalia* or dry docks, built by Hermodorus of Salamis, in the second half of the second century BC).

2.30. Welsh, *op. cit.* (2.15), 116.

2.31. J.G. Landels, *Engineering in the Ancient World* (London, 1978), 66.

2.32. Cotterell and Kamminga, *op. cit.* (2.7), 220–1.

2.33. Dijksterhuis, *op. cit.* (2.6), 18.

2.34. Landels, *op. cit.* (2.31), 191.

2.35. G.J. Tee, Letter, *New Scientist*, 10 June 1989: 49.

2.36. G.E.R. Lloyd, 'Hellenistic Science: Its Application in Peace and War', ch. 9, Part I, *The Hellenistic World*, of the *The Cambridge Ancient History*, eds F.W. Wallbank *et. al.* (Cambridge, 1984), 8: 335.

2.37. W. Knorr, *The Ancient Tradition of Geometrical Problems* (Boston, 1986), 190–1 (use of proof), 194 (solutions).

3.1. *Marc., op. cit.* (1.4), 14.7–9, 99.

3.2. Heath, *op. cit.* (1.3), xix; Pappus, *La Collection mathématique*, Book 8, sect. 11, prop. 10, trans. P. ver Eecke, 2nd edn (Paris, 1982), 836–7.

3.3. *Aristotelis Physicorum libros quattuor posteriores commentaria*, ed. H. Diels (Berlin, 1895), 1110, trans. A.G. Drachmann, in 'Fragments from Archimedes in Heron's Mechanics', *Centaurus*, 1963, 8: 9–146, 143.

3.4. Dijksterhuis, *op. cit.* (2.6), 16, and n. 7 (Tzetzes, *Chil.* 3, *Hist.* 66, 62) and n. 8 (Tzetzes, *Chil.* 2, *Hist.* 35, 130). The first mentions the *trispaston* and the other the *charistion.* Drachmann, 1956–8 (2.9), 5, 278–81, 280 (Tzetzes, *Chil.* 2, *Hist.* 35: 128, *Chil.* 2, *Hist.* 66, 60). Note slight differences in line numbers. Drachmann underestimated Tzetzes' capacity for misquotation.

3.5. Drachmann, 1956–8 (2.9), 280. Drachmann, 1973 (2.9), 50. The Golden Rule of Mechanics is that the effort in moving a given load is inverse to the distance over which the effort has to be exerted. It applies not only to the steelyard and the lever, but also to the pulley, the rope and drum, the gearwheel and the endless screw.

3.6. Dijksterhuis, *op. cit.* (2.6), 18.

3.7. *Ibid.*, 17–18. There is some confusion in his last sentence, because until then Dijksterhuis had been discussing the origins of the story of moving the earth, not moving the Great Ship; the resultant belief matches the attitude in Classical and later times to Archimedes.

3.8. *Ibid.*, 363.

3.9. H. Arendt, *The Human Condition* (New York, 1958), 13.

3.10. *Ibid.*, 238 (Greek original transliterated as *dos moi pou sto*); cf. Dijksterhuis, *op. cit.* (2.6), 15 (translation) and 16 (see Pappus, *op. cit.* (3.2), 8.11.10, 836–7).

3.11. Arendt, *op. cit.* (3.9), ch. 6, 225. This usage may be common in German-speaking countries because György Dalos remarked that 'Our now-disorganised world really does miss an Archimedean point', *The Guardian*, 7 November 1991: 29.

3.12. Arendt, *op. cit.* (3.9), 295. Miss Arendt's metaphors do rather obscure her meaning to an Archimedean reader.

3.13. *The Guardian*, 12 May 1992.

4.1. Pappus, *op. cit.* (3.2), 8.3, 813 (Carpus), 813 footnote 3 (probably second century AD). Firmicus, *op. cit.* (1.13), 1–2 (fourth century AD Sicilian), 11, Proemium, para. 5 (sphere); 210, Book 6, para. 30 (loss of book on sphere).

4.2. M.T. Cicero, *De re publica*, 1.14.21–3, trans. C.W. Keyes (London, 1928), 41, 43.

4.3. *Marc.*, *op. cit.* (1.4), 19, 105.

4.4. M.T. Cicero, *Tusculanarum disputationum*, 1.25.63–6, trans. J.E. King (London, 1966), 1: 73, 75; M.T. Cicero, *De natura deorum*, 2.34.87–8, trans. H. Rackman (London, 1933), 208–9.

4.5. Ovid, *Fasti*, 6.263–83, trans. J.G. Frazer (London, 1976), 341; Claudian, *Epigrams*, 51.68, trans. M. Platnauer, two volumes (London, 1922), 2: 278–80, line 6.

4.6. Lactantius, *Divinae institutiones*, 2.5.18, rec. S. Brandt, *Opera*, Part I (Vienna, 1890); Firmicus, *op. cit.* (1.13), 11, 303, nn. 3 and 210.

4.7. Martianus Capella, *De nuptiis philologiae et Mercurii libri*, Lib. 2, 10–15, ed. A. Dick (Leipzig, 1925), 78; Macrobius, *Somnium Scipionis*, 1.19.2, ed. J. Willis, two volumes (Leipzig, 1963), 2: 73.

4.8. *The Letters of Cassiodorus*, 1.45 and 1.46, trans. T. Hodgkin (London, 1886), 168–70, 1.45 (order for water-clocks and mention of Archimedes). A second letter recorded that both clocks had been delivered to the Burgundian King (Cassiodorus, 1.46). Chadwick doubted whether Boethius possessed the practical technical competence to build a water-clock (H. Chadwick, *Boethius: The Consolations of Music, Logic, Theology and Philosophy* 1st edn 1981 (Clarendon paperbacks, Oxford, 1990), 23.) This is misguided. Boethius was called upon to design the waterclock; that done, a skilled craftsman should be able to undertake the detailed design, the detailing and the construction. Neither Hill nor Sleeswyk refers to Cassiodorus' silence about the originator of the water-clock. Did Gerbert know of this reference in Cassiodorus? According to E. Grant, *Physical Sciences in the Middle Ages* (Cambridge, 1977), 14: 'Gerbert not only explained how to construct a sphere to represent the heavens, but actually made one which simulated the motions of the constellations, using wires fixed on the surface of a sphere to outline the stellar configurations.' Claudian did not claim to have seen a planetarium. Did the two devices made by Archimedes disappear in the great fire of Rome in the tenth year of Nero's reign? E. Gibbon, *The Decline and Fall of the Roman Empire*, six volumes, Everyman edition (London, 1911), 2: 14 (Tacitus, *Annals*, 15.38–44).

4.9. Sextus Empiricus, *Adversos physicos*, 1.115, trans. R.G. Bury, *Against the Physicists* (London, 1936), 3: 63.

4.10. The original may give bronze, but the translator has chosen the word brass instead.

4.11. Heath, *op. cit.* (1.3), xxi (Hultsch); Eecke, *op. cit.* (3.2), 812.

4.12. Eecke, *ibid.*, 813, footnote 6.

4.13. *On the Construction of Water-Clocks*, ed. and trans. D.R. Hill, Turner and Devereux, Occasional Paper No. 4 (London, 1976). The first twenty-nine-and-a-half folios of the Paris manuscript contain a treatise on the sphere, but tantalisingly Hill made no reference to this in his publication. Hill, *ibid.*, intro., 6.

4.14. *Ibid.*, 8.

4.15. *Ibid.*, 7 (Arab authors), 9 (apparatus).

4.16. Sleeswyk, *op. cit.* (2.20), 34 (Vitruvius on Ktesibios), 29 (Procopius).

4.17. J. Needham, *Science and Civilisation in China*, seven volumes (Cambridge, 1954), 1: 230. D.R. Hill, 'The Toledo Water-Clocks', *Hist. Tech.*, 1994, 16: 62–71.

4.18. Field and Wright and others have reviewed the evidence of Greek and Byzantine knowledge of gearing. J.V. Field and M.T. Wright, *Early Gearing* (London, 1985); J.V. Field and M.T. Wright, 'Gears from the Byzantines: A Portable Sundial with Calendrical Gearing', *Annals of Science*, 1985, 42: 87–138; A.G. Bromley, 'Notes on the Antikythera Mechanism', *Centaurus*, 1986, 29: 5–27; M.T. Wright, 'Rational and Irrational Reconstruction: The London Sundial Calendar and the Early History of Geared Mechanisms', *Hist. Tech.*, 1990, 12: 65–102.

4.19. Field and Wright, *op. cit.* (4.18). D.R. Hill, 'Al-Biruni's Mechanical Calendar', *Annals of Science*, 1985, 42: 139–63.

5.1. Diodorus, *op. cit.* (1.8) 1.34, 1: 199.

5.2. Vitruvius, *The Ten Books on Architecture* 10.6.1–4, trans. M.H. Morgan (New York, 1960), 295–7, 297.

5.3. Landels, *op. cit.* (2.30), 61–2.

5.4. J.P. Oleson, *Greek and Roman Mechanical Water-Lifting Devices: The History of a Technology*, Journal of the Classical Association supplementary volume 16 (London, 1964), 296; Landels, *op. cit.* (5.3), 61–2; Ann Woods, 'Mining', in *The Roman World*, ed. J. Wacher, two volumes (London and New York, 1987), 2: 610–34, 620.

5.5. Oleson, *op. cit.* (5.4), see especially 291– 300.

5.6. *Ibid.*, 61.

5.7. Diodorus, *op. cit.* (1.8), 1.34, 1: 199; 5.37.3–4, 3: 196. Both references state that Archimedes was the inventor. Forbes is wrong in claiming that Diodorus said that screw was in use before Archimedes' time. See R.J. Forbes, 'Hydraulic Engineering', in *A History of Technology*, eds C. Singer, E.J. Holmyard, A.R. Hall and Trevor Williams (Oxford, 1956), 2: 663–94, 676–7.

5.8. Oleson, *op. cit.* (5.4), 299–301.

5.9. Landels, *op. cit.* (2.31), 66. Sarton, *op. cit.* (2.2), 2: 122–5: description from Athenaios, *Deipnosophists.*, 5.203–9, *op. cit.* (2.2), 2: 421–47 (pump bilge-water).

5.10. Oleson, *op. cit.* (5.4), 291–2.

5.11. *Ibid.*, 293.

5.12. Vitruvius, *op. cit.* (5.2), 6.1–4, 295–7; 7.1–5, 297–8.

5.13. *Ibid.*, 297. See also Philo Judaeus, 'De confusione linguarum', ch. 38 in *Philonis Alexandrini opera quae supersunt*, ed. P. Wendland (Berlin, 1897), 2, 236.

5.14. Oleson, *op. cit.* (5.4), 298.

5.15. British Museum No. 37563.

5.16. Drachmann, 1973 (2.9). His claim that Archimedes used the crank is wholly unconvincing.

5.17. Ahmad Y. al-Hassan and D.R. Hill, *Islamic Technology* (Paris, 1992), 243. D.R. Hill, 'Mechanical Engineering in the Medieval Near East', *Scientific American*, May 1991, 264: 64–9, 66 (Arabs had crank with connecting rods), 68 (invented crank).

5.18. J. Needham, 'Science and China's Influence in the World', in *The Legacy of China*, ed. R. Dawson (Oxford, 1964), 234–308, 278. Lynn White Jr, 'Theophilus Redivivus', *Tech. and Cult.*, 1964, 5: 224–33, 233 (crank in West first found in Utrecht Psalter, second quarter of ninth century), 233 (Theophilus takes use of crank for granted).

5.19. D.J. de Solla Price, 'Gears from the Greeks', *Trans. Amer. Phil. Soc.*, 1964, 64: 7, 20 (ref. to Drachmann), 40 (crank, as I suppose), 41 (presumed crank handle).

5.20. K.D. White, *Greek and Roman Technology* (London, 1984), 50.

5.21. *Ibid.*, 173, quoting J.F. Healey, *Mining and Metallurgy in the Greek and Roman World* (London, 1978), 95. Woods, *op. cit.* (5.4), 620.

5.22. Attributed to Palmer, but taken from Rickard. See Healy, *op. cit.* (5.24), 96–7; Woods, *op. cit.* (5.4), 620, Fig. 24.3; White, *op. cit.* (5.21), 23, Fig. 12; R.E. Palmer, 'Notes on Some Ancient Mining Equipments and Systems', *Trans. Inst. Mining and Metallurgy*, 1926–7; 36: 299–310, 301.

5.23. Palmer, *ibid.* (5.22), 328–33, 329.

5.24. T.A. Rickard, 'With Geologists in Spain', *Engineering and Mining Journal*, 4 June 1927, 123: 917–23, 922; 2 July 1927, 124: 13–18; 16 July 1927, 124: 91–5 (original of Centenillo drawing by Haselden).

5.25. Oleson, *op. cit.* (5.4), 400, 289–90.

5.26. *Ibid.*, 299.

5.27. *Ibid.*, 230.

5.28. *Ibid.*, 291.

5.29. Even the credulous Mirabella, who believed that Archimedes had invented artillery, realized that so much had been attributed to Archimedes that the credibility of this attribution to him was diminished. V.E. Alagova Mirabella, 'Vita d'Archimede Siracusano', in *Dell'antiche Siracuse* (Naples, 1613), book 2, 106–10, 109.

5.30. *Les Dix Livres d'architecture de Vitruve*, trans. C. Perrault (Brussels, 1979), 316, footnote 4.

5.31. A.G. Drachmann, *The Mechanical Technology of Greek and Roman Antiquity* (Copenhagen, 1963), 152–4.

5.32. Vitruvius, *op. cit.* (5.2), 10.4.1–2, 203– 4.
5.33. Oleson, *op. cit.* (5.4), 293.
5.34. *Ibid.*, 291.
5.35. *Ibid.*, 294.
5.36. *Ibid.*, 292.
5.37. Heath, *op. cit.* (1.3), 151–5, 151.
5.38. Diodorus, *op. cit.* (1.8), 5.37.3, 3: 199.
5.39. Alternatively, Archimedes intended it as a bilge-pump for the *Siracusia.* Sarton, *op. cit.* (5.11), 2: 421–47. When Hieron presented the Ship to Ptolemy, the Egyptians adopted it for irrigation. This is less convincing.
5.40. Oleson, *op. cit.* (5.4), 299; Landels, *op. cit.* (2.31), 61–2; Woods, *op. cit.* (5.4), 2: 610–34, 620; Healy, *op. cit.* (5.24), 95.
5.41. Vitruvius, *op. cit.* (5.2), 7, intro, para. 14, 199.
5.42. Oleson, *op. cit.* (5.4), 298.
5.43. *Ibid.*, 406. The formulae for arrow-shooters, rock-throwers and catapults to be operated at selected ranges are the military equivalents.
5.44. *Ibid.*, 406.
5.45. *Ibid.*, 294.
5.46. *Ibid.*, 294 (rarity).
5.47. Forbes, 'Hydraulic Engineering', in Singer *et al.*, *op. cit.* (5.7), 2: 676–7. Oleson, *op. cit.* (5.4), 294.
5.48. B. Gille, *The Renaissance Engineers*, trans. unknown (London, 1966), 64.
5.49. *Ibid.*, 155. Vitruvius, *De architectura*, ed. Fra Giocondo (Verona, 1511). Archimedes has not been lucky in the accuracy of his illustrators. See D.L. Simms, *An Iconography of Pseudo-Archimedes* (in preparation).
5.50. D.J. Bryden and D.L. Simms, 'Archimedes and the Opticians of London', *Bull. Sci. Instr. Soc.*, 1992, 35: 11–14.
5.51. Hassan and Hill, *op. cit.* (5.19), 243 (Fig. 9. 7b). Hill, *op. cit.* (5.20) (no dates given).
5.52. Landels, *op. cit.* (2.30), 62–3.
5.53. H. Hodge, *Technology in the Ancient World* (Harmondsworth, 1971), 185.
5.54. Oleson, *op. cit.* (5.4), 290.
6.1. Polybius, *Rise, op. cit.*, (1.10), 1.16, 57.
6.2. *Ibid.*, 3.76, 244; Livy, *op. cit.* (1.6), 22.37, 135–6 (claimed that the supplies were freely sent).
6.3. Polybius, *op. cit.* (1.10) 8.7, 368.
6.4. *Ibid.*, 7.2, 353.
6.5. P. Green, *Alexander to Actium: The Hellenistic Age* (London, 1990), 226 (n.b. he was the grandson, *not* the son).
6.6. Polybius, *Rise, op. cit.* (1.10), 7.3, 354.
6.7. Polybius, *Histories, op. cit.* (1.10), 8.3–7, 3: 451–63 (attack by sea and land); 8.6.37, 537–9 (fragment of capture of the city). Livy, *op. cit.* (1.6), 24.33–4, 271–4; *Marc., op. cit.* (1.4), 14–20, 98–105.
6.8. F.W. Walbank, *Commentary on Polybius*, three volumes (Oxford, 1967), 1: 2, n. 4.
6.9. E.W. Marsden, *Greek and Roman Artillery: Technical Treatises* (Oxford, 1971), 6.
6.10. Polybius, *Rise, op. cit.* (1.10), 8.3–7, 364–8 (attack by sea and land); *Histories, op. cit.* (1.10), 8.6.37, 3: 537–9 (fragment of capture of the city); Walbank, *op. cit.* (6.8), 2: 72–7. Marsden, *op. cit.* (6.9), 92 (description of the *tollenones* or swing-beam with iron grapnel omitted the fact that they were counter-balanced by a lead weight; Livy has the best account of this weapon). Cf. Livy, *op. cit.* (1.6), 24.34.10, 273. Cullham has argued that Plutarch's account differs from those in Polybius and Livy. The differences she identified lie in the details and do not affect the general agreement of the three versions. Phyllis Cullham, 'Plutarch on the Roman Siege of Syracuse: The Primacy of Science over Technology', in *Plutarco e la Scienza*, ed. I. Gallo (Genoa, 1992), 179–97, 179–80.
6.11. Livy, *op. cit.* (1.6), 24.33, 272 (confident of their ability); *Marc., op. cit.* (1.4), 14, 98 (confident in the size of his armament to overawe the Syracusans). Polybius, *Rise, op. cit.* (1.10), 8.3, 365.
6.12. Polybius, *Histories, op. cit.* (1.10), 5.88, 3: 217.

6.13. W.W. Tarn, *Hellenistic Military and Naval Developments* (Cambridge, 1930), 102 (Dionysius of Syracuse). E.W. Marsden, *Greek and Roman Artillery: Historical Development* (Oxford, 1969), 48, 54–5 (against Carthaginians at Motya). Diodorus, *op. cit.* (1.8), 14.50.4, 6: 153.

6.14. Polybius, *Rise, op. cit.* (1.10), 8.5, 366. Marsden converted this version by Polybius into an elegant account of how a hypothetical attack on the Euryalos fort would be met: Marsden, *op. cit.* (6.13), 119.

6.15. Polybius, *Rise, op. cit.* (1.10), 8.3, 364.

6.16. *Ibid.*, 8.5–6, 366–7.

6.17. Marsden, *op. cit.* (6.13), 108 (not, as Marsden suggests, that Marcellus had neutralized some of the batteries).

6.18. Polybius, *Rise, op. cit.* (1.11), 8.5–6, 366–7.

6.19. *Ibid.*, 8.7, 367–8.

6.20. *Ibid.*

6.21. A.W. Lawrence, *Greek Aims in Fortification* (Oxford, 1979), 427 (Lawrence concurred with Philon).

6.22. Josephus, *The Jewish War*, 6, 402, trans. G.A. Williamson (Harmondsworth, 1970), 358.

6.23. Polybius, *op. cit.* (1.10), 8.5–6, 366–7; 8.7, 367–8.

6.24. Walbank, *op. cit.* (6.8), 2: 75, n. 9.

6.25. Marsden, *op. cit.* (6.13), 50–1 (no evidence before then; particularly noted no mention in Thucydides, who was interested in military machines, or in the corresponding part of Diodorus).

6.26. *Ibid.*, 57.

6.27. *Ibid.*, 79–80.

6.28. *Ibid.*, 91, n. 5.

6.29. *Ibid.*, 86 (Marsden reasonably thought he could do better).

6.30. *Observer Magazine*, 10 November 1974, 86.

6.31. Marsden, *op. cit.*, 98, 107 (Demetrius' siege of Rhodes).

6.32. W. Soedel and V. Foley, 'Ancient Catapults', *Scientific American*, 1979, 240: 120–8.

6.33. Marsden, *op. cit.* (6.13), 62; Marsden, *op. cit.* (6.9), 107–9.

6.34. K.D. White, *op. cit.* (5.20), Appendix 15, Development of the catapult: 217 (holes so punched as to make possible very fine tensioning or 'tuning' of the sinew-ropes of the torsion-type weapons); 218 ('key dimension, they [the Greek military engineers] found, was the size of the skeins, which was determined by the diameter of the holes that received them. All the other components were proportionate to this diameter, ... established by altering the sizes and testing the results, ... formula ... arrow-shooting type simple ... diameter ... 1/9 of the length of the arrow ... stone-thrower was more complicated ... some workshops ... had lists worked out by a tame mathematician').

6.35. Marsden, *op. cit.* (6.13), 109–10, 122; *Marc.*, *op. cit.* (1.4), 16.2, 101.

6.36. A. Kircher, *Ars magna lucis et umbrae* (Rome, 1646), 876. W.E. Knowles Middleton, 'Archimedes, Kircher, Buffon and the Burning Mirrors', *Isis*, 1961, 52: 533–43, 536 and n. 17.

6.37. Lawrence, *op. cit.* (6.21), 278.

6.38. Marsden, *op. cit.* (6.13), 108.

6.39. *Ibid.*, 92.

6.40. Walbank, *op. cit* (6.8), n. 6, 2: 75; S. Price, 'The History of the Hellenistic Period', in *The Oxford History of the Classical World*, ed. J. Boardman, J. Griffin and O. Murray, two volumes, *Greece and the Hellenistic World* (Oxford, 1988), 1: 309–31, 323 (photograph by A.W. Lawrence).

6.41. Marsden, *op. cit.* (6.9), 51–2 (Plataeans against Peloponnesians, 429 BC; Conon against Spartans).

6.42. Kircher, *op. cit.* (6.36), 876. Landels, *op. cit.* (2.31), 98 (a swivel mounting with bearings at each side).

6.43. Soedel and Foley, *op. cit.* (6.31), 120. Landels, *op. cit.* (2.31), 95.

6.44. A rare example of lead actually harming the Romans. See D.L. Simms, 'Lead in History, Lead and History, Leaden History', *Science of the Total Environment*, 1984, 37: 259–66.

6.45. Landels, *op. cit.* (2.31), 96–7 (who should know better than Archimedes?).

6.46. *Ibid.*, 97–8.

6.47. *Marc., op. cit.* (1.4), 15, 100.

6.48. Josephus, *op. cit.* (6.22), 3.243, 200; 3.171, 193.

6.49. D.L. Simms, 'The Burning Mirrors of Syracuse', *T. & C.*, 1977, 20: 1–17, 7, D.L. Simms, 'Galen on Archimedes: Burning Mirror or Burning Pitch?', *T. & C.*, 1991, 32: 91–6.

6.50. *Pappus of Alexandria. Book 7 of the Collection*, two volumes, intro., text, trans. A. Jones (New York, 1986), 1: 38. (Tzetzes, Dion and Diodorus record the story, and many along with them about Archimedes, Anthemius the paradoxographer, Heron and Philon, Pappus and every writer on mechanics, in whom I have read about reflective ignitions. Dio, *Roman History*, trans. E. Carey (London, 1914), 2: 171–3; Zonaras, *Epitome, 14.3* (quoting from Dio Cassius). Probably by then only excerpts of Polybius available; Plutarch was extant, only book in Tzetzes' library. Zonaras included material from him. F. Millar, *A Study of Cassius Dio* (Oxford, 1964), 56, n. 4.

6.51. Gibbon, *op. cit.* (4.8), 1: 176, n. 1 (little dependence is to be had on the authority of a moderate Greek [Zonaras] so grossly ignorant of the history of the third century that he creates several imaginary emperors and confounds those who really existed). Gibbon, *loc. cit.*, 4: 364 (source of this idle fable, the *Chiliads* of John Tzetzes).

6.52. J. Caesar, *The Civil War*, trans. A.G. Peskett (London, 1914), 2.1.11, 139; 2.1.14, 143. Josephus, *op. cit.* (6.23), 3.129, 198.

6.53. Simms (1977), *op. cit.* (6.49), 1–17.

6.54. A.A. Mills and R. Clift, 'Reflections on the "Burning Mirrors of Archimedes" ', *Eur. J. Phys.*, 1992, 13: 268–79.

6.55. Carey, *op. cit.* (6.50), 2: 171–3; Zonaras 9.4 (kindled a great flame, directed upon the ships that lay at anchor, until he consumed them all). Carey, *ibid.*, 2: 171–2; Tzetzes, *Chil.* 2: 109–28 ('When Marcellus withdrew them a bowshot off . . . its [sun's] noontide beam . . . So after that . . . reduced them [the ships] to ashes a bowshot off'). Simms (1977), *op. cit.* (6.49), 8, n. 33.

6.56. Simms, *ibid.* (incorporated from an earlier passage in Anthemius).

6.57. Marsden, *op. cit.* (6.13), 12.

6.58. Buffon's selection of a bowshot as 150 *pied* (Parisian feet) was quite arbitrary. That choice was part of his unusual decision to carry out his experiments before examining the historical evidence. See D.L. Simms, 'Those Burning Mirrors of Buffon' (in preparation).

6.59. Mills and Croft, *op. cit.* (6.54), 276, 278.

6.60. The period may have been much longer. Gelon was, unusually, joint tyrant with his father, so that Hieron could have relinquished the reigns of office and had to take them up again as a very old man when his son died.

6.61. Polybius, *Rise, op. cit.* (1.10), 8.7, 367–8 (artillery both in the volume and the velocity of its fire, as was to be expected when Hieron had provided the supplies and Archimedes designed the various engines). Livy, *op. cit.* (1.6), 24.34.13, 273; *Marc., op. cit.* (1.4), 14–7.9, 98.

6.62. Marsden, *op. cit.* (6.13), 62.

6.63. Marsden, *op. cit.* (6.9), 122, 125 (the corridor wall at Selinus – same purpose).

6.64. Marsden, *op. cit.* (6.13), 115 (Philo). Walbank, *op. cit.* (6.8), 2: 75, n. 6 (Heron).

6.65. Walbank, *ibid.*, 75, n. 9. Polybius, *Histories, op. cit.* (1.10), 21.27.4, 5: 295–7.

6.66. Marsden, *op. cit.* (6.13), 3, 4 and 189. Marsden, *op. cit.* (6.9), 2. Lawrence, *op. cit.* (6.21), 69–110 (comprehensive account *Polioketika*).

6.67. Vitruvius, *op. cit.* (5.2), 10, 11, 12, 13, 303–11, 311 (Diades); 14, 15, 16, 311–19, 311 (drawn from own experience).

6.68. F.R. Vegetius, *De re militari* (London, 1797). E.A. Thompson, *A Roman Reformer and Inventor* (Oxford, 1952).

6.69. B.C. Hacker, 'Greek Catapults and Catapult Technology: Science, Technology and War in the Ancient World', *T. & C.*, 1968, 9: 34–50. There is nothing here to support Drachmann's claim that 'Archimedes designed a very powerful crane for the defence of Syracuse . . . After his death it fell into disuse': A.G. Drachmann, 'A Note on Ancient Cranes', in Singer *et al.*, *op. cit.* (5.7), 658– 62, 658.

6.70. Marsden, *op. cit.* (6.13), 108.

6.71. A.W. Lawrence, 'Archimedes and the Design of the Euryalos Fort', *J. Hellenic Studies*, 1946, 66: 99–107 (opinion based upon his experience during World War II).

6.72. Landels, *op. cit.* (2.31), 94–5; White, *op. cit.* (5.20), 47–8.

6.73. Polybius, *Rise, op. cit.* (1.10), 8.3, 365.

6.74. Livy, *op. cit.* (1.6), 26.47, 417.

6.75. *Ibid.*, 25.31, 338.

6.76. White, *op. cit.* (5.20), 218.

6.77. J. Caesar, *The Conquest of Gaul*, trans. S.A. Handford (Harmondsworth, 1951), 220, 223, 239 (Alesia, 52 BC). Josephus, *op. cit.* (6.22), 3.171, 193–4 (Jotapata); 5.21, 5.46, 277; 5.269, 297–8 (Jerusalem); 7.304, 384 (Masada).

6.78. Caesar, *op. cit.* (6.52), 2.1, 125; 2.1.11, 139.

6.79. That circle of Scipio Aemilianus (185–129 BC) must have been familiar with artillery; Scipio's adopted grandfather was Scipio Africanus, the victor of Zama, which ended the Second Punic War.

6.80. Josephus, *op. cit.* (6.22), 3.148, 3.171, 192–4.

6.81. Procopius of Caesarea, *History of the Wars* Book V, ch. 22.18–22, trans. H.B. Dewing, seven volumes (London 1919), 1: 203–17.

6.82. Cassius Dio, *op. cit.* (6.50), 2: 171–3 (Zonaras, 9.4; Tzetzes, *Chil.*, 2, 109–28).

6.83. Athenaios, *Deipnosophists, op. cit.* (2.2), 5.206d–208f, 2: 435–45, (stonethrower); 14.634b, 6: 421 (Polybius' account). Proclus, *op. cit.* (2.4), 41.13, 33 (Polybius on weapons Archimedes is credited with devising).

6.84. Athenaios, *op. cit.* (2.2), 5.203–9, 2: 421–47.

6.85. Livy, *op. cit.* (1.6), 24.33–4, 271–4; 25.23–31, 326–38; Silius Italicus, *op. cit.* (2.3), 14.300–15, 2: 295: Pliny, *op. cit.* (1.13), 7.37.125, 589–91; Quintilian, *Instituto orata* 1.10.48, trans. H.E. Butler, four volumes (London, 1921–2), 1: 183 (if Quintilian (AD 35–100) said so, then every schoolboy knew). *Ausonius*, trans. H.G. Evelyn White, four volumes (London, 1919), 1: 247, lines 303–4.

6.86. Marsden, *op. cit.* (6.10), 264.

6.87. R.R. Bolgar, *The Classical Heritage and Its Beneficiaries* (Cambridge, 1954), 193.

6.88. Simms, *op. cit.* (1.16), 195–210.

6.89. F. Petrarch, *De viris illustribus* and *Rerum memorandarum libri*; see M. Clagett, 'Biographical Accounts of Archimedes in the Middle Ages', Appendix 3 in *Archimedes in the Middle Ages* (Philadelphia, 1978), 3: 1329–41, 1336, 1338.

6.90. Simms, *op. cit.* (1.16), 67–79 (an additional reference: W.T., 'The Compleat Gunner in Three Parts, Translated out of Casimir, Diego, Ussano and Hexam etc.) Being Book III Part 1 of' *Military and Maritime Discipline in Three Books* (London, 1672), 1 (Sig. 3B1^{r}: 'We will not dispute of the first invention of Guns, that is whether it came from Archimedes as the *Italians* do report'); Gibbon did know the passage in Petrarch: 'Miscelleana', in *The English Essays of Edward Gibbon*, ed. Patricia B. Craddock (Oxford, 1972), 330. Petrarch speaks of the use of gunpowder *nuper rara. nunc communis.* de Remed. utriusque L(ibre) 1 Dial(ogue) 99. Gibbon thus refrained from commenting upon it; see Simms, *op. cit.* (1.16), 77, n. 20.

6.91. R. Valturius, *De re militari* (Paris, 1532), 265: the first edition (Verona, 1483) is unpaginated.

6.92. Gibbon, *op. cit.* (4.8), 4: 254 and 254, n. 2 (I have seen an ingenious model, of an *onagri*, contrived and executed by General Melville, which imitates or surpasses the art of antiquity).

6.93. A. Koyré, 'Commentary', in *Scientific Change*, ed. A.C. Crombie (London, 1963), 847–61, 855, n. 1.

6.94. Lawrence, *op. cit.* (6.71), 99–107; M. Clagett, 'Archimedes', in *DSB* (New York, 1970), 1: 213–31, 214.

6.95. Dijksterhuis, *op. cit.* (2.6), 28.

6.96. Marsden; *op. cit.* (6.13), 125.

6.97. Lawrence, *op. cit.* (6.71), 99–107.

6.98. M.I.Finley, 'Technical Innovation', in *Economy and Society in Ancient Greece*, ed. B.D. Shaw and R.P. Saller (Harmondsworth, 1983), 181 (Archimedes' practical inventions, I hasten to add, were military and made only under the extraordinary and irresistible stimulus of the siege of his native Syracuse by the Romans). Green, *op. cit.* (6.5), 471 (it took the extraordinary stimulus of the siege of Syracuse to make him apply his knowledge to problems of siege warfare).

6.99. B. Russell, *Human Society in Ethics and Politics* (London, 1954), 218.

7.1. *Marc., op. cit.* (1.4), 19.3, 102.

7.2. *Ibid.*, 14, 98.

7.3. *Ibid.*, 14, 98–9.

7.4. Similar passage, with the addition of the name of Menaechmus, in 'Table Talk, 8.2.718', in Plutarch's *Moralia*, 17 volumes, trans. E.L. Minar, F.H. Sandbach and W.C. Helmbold (London, 1961), 9: 121–3 (it was for this reason that Plato himself reproached Eudoxus and Archytas and Menaechmus for setting out to remove the problem of doubling the cube into the realm of instruments and mechanical devices, as if they were trying to find two mean proportionals not by the use of reason but in whatever way would work. In this way, he thought the advantage of geometry was dissipated and destroyed, since it slipped back into the realms of sense-perception instead of soaring upward and laying hold of the internal and immaterial images in the presence of which God is always God).

7.5. D.A. Russell, *Plutarch* (London, 1972), 147. R. Pfeiffer, *A History of Classical Scholarship: 1300–1850* (Oxford, 1976), 29.

7.6. Clagett, *op. cit.* (6.89), vol. 3, pt 4, Appendix 3, 1329–41. G. Sarton, *The Appreciation of Ancient and Medieval Science during the Renaissance (1450–1600)* (Philadelphia, 1955), 214–15, n. 19.

7.7. Gille, *op. cit.* (5.51), 10, 83; Gille, 'Machines', in Singer *et al., op. cit.* (5.7), 629–62, 632–3.

7.8. Guido Ubaldo, *Le Mechaniche* (Venice, 1581), quoted by R. Olson, *Science Deified and Science Defied* (Berkeley and Los Angeles, 1982), 227.

7.9. *The Essays of Michel de Montaigne*, trans. and ed. M.A. Screech (Harmondsworth, 1987), 152.

7.10. Sarton, *op. cit.* (2.2), 2: 70. A.R. Hall, 'Military Technology', in Singer *et al., op. cit.* (5.7), 2: 695–730, 714. R.J. Forbes, 'Power', in *loc. cit.* (5.7), 589–622, 604. L. Wolpert, *The Unnatural Nature of Science* (London, 1992), 40–1 (unnatural nature of checking one's references?). J.D. Bernal, *Science in History* (London, 1954), 158.

7.11. Oleson, *op. cit.* (5.4), 292, 294, 298. See Vitruvius, *op. cit.* (5.2), book 7, intro., para. 14, 199.

7.12. Hacker, *op. cit.* (6.69), 42–3 (Dionysius), 47 (competition/attitude), 47–8 (diversions).

7.13. Green, *op. cit.* (6.5), 458, 462 (there is a great deal of evidence that he simply inherited the intellectual tradition of Plato and Aristotle), 470, 471 (cf. conscious formulation by Archimedes . . . it took the extraordinary stimulus of the siege of Syracuse to make him apply his knowledge to problems of siege warfare . . . When Syracuse fell, Archimedes was absorbed in a mathematical proposition, oblivious to the legionary who cut him down. He died as he would have wanted).

7.14. Finley, *op. cit.* (6.98), 181. Green, *op. cit.* (6.5), 458, 462, 470, 471. Cullham, *op. cit.* (6.12), 190.

7.15. Heath, *op. cit.* (1.3), xvi. Readers may appreciate the irony of Heath's working life spent in the UK Civil Service; he was Joint Permanent Secretary to the Treasury during World War I, retiring as Comptroller, National Debt Office.

7.16. Dijksterhuis, *op. cit.* (2.6), 25; 47–9 (elements of mechanics; water-clocks), 13, 21. The two later comments distance Dijksterhuis from Plutarch, though Knorr suggested Dijksterhuis was silent. Dijksterhuis, *loc. cit.*, 441. Cullham argued the reverse. Cullham, *op. cit.* (6.10), 181–2.

7.17. G. Santillana, *The Origins of Scientific Thought* (New York, 1961), 239–40. E.J. Stillman Drake, *Cause, Experiment and Science* (Chicago, 1981), 35–6. G.E.R. Lloyd, *Greek Science after Aristotle*, two volumes (London, 1973), 2: 95; Lloyd, *op. cit.* (2.36), 337. Cullham, *op. cit.* (6.10), 179–97. A. Georgiadou, 'The Corruption of Geometry and the Problems of Two Mean Proportionals', in *ibid.*, 147–64, p. 147. K.D. White, 'The Base Mechanic Arts? Some Thoughts on the Contribution of Science (Pure and Applied) to the Culture of the Hellenistic Age', in *Hellenistic History and Culture*, ed. P. Green (Berkeley, 1993), 9: 211–20, 219. Clagett, *DSB* (6.94), 1: 213–31, 214a. A.G. Drachmann, 'Archimedes and the

Science of Physics', *Centaurus* 1968, 12: 1–11, pp. 2, 10, 11; Landels, *op. cit.* (2.31), 187; Cotterell and Kamminga, *op. cit.* (2.7), 11.

7.18. W.C. Helmbold and E.N. O'Neil, *Plutarch's Quotations*, Philological Monographs (Oxford, 1959), ix (preface), 63–4.

7.19. *Plutarch's Lives. Cicero*, intro., trans., comm. J.L. Moles (Warminster, 1988), 39, 43–4.

7.20. Scott-Kilvert, *op. cit.* (1.4), intro., 8.12.

7.21. Green, *op. cit.* (6.5), xix.

7.22. *Marc., op. cit.* (1.4), 19, 104–5. Livy, *op. cit.* (1.6), 25.31. 338.

7.23. Simms, *op. cit.* (1.14), 283.

7.24. *Marc., op. cit.* (1.4), 17, 103. *Plutarch's Moralia* (7.4), 786C, 10: 93. *Ibid.*, 1093, 64–5 ('how exquisite ... the pleasure ... on the quadrant'); 69 ('raptures ... of Eudoxus and Archimedes and Hippocrates ... so strong is its [mathematics'] spell'). Cullham noted the emphasis on the pleasure of mathematics in contrast to that from its use. Cullham, *op. cit.* (6.10), 181.

7.25. *Ibid., Moralia* (7.4), 1094C, 14: 15–149, 67. Vitruvius, *op. cit.* (5.2), 9, 9–12, 253–4.

7.26. Heath, *op. cit.* (1.3), 18.

7.27. White, *op. cit.* (7.17), 219.

7.28. *Marc., op. cit.* (1.4), 19, 105.

7.29. Heath, *op. cit.* (1.3), xxxvi–xxxviii; Dijksterhuis, *op. cit.* (2.6), 47–9.

7.30. Knorr, *op. cit.* (2.37), 294. Heracleides, his biographer, the most likely candidate, can hardly have done so, since Eutocius, who knew the biography, was not familiar with many of them.

7.31. Heath, *The Method, op. cit.* (1.3), 13; see also Lloyd, *op. cit.* (7.17), 46; Lloyd, *op. cit.* (2.36), 321–52, 352; Santillana, *op. cit.* (7.17), 239– 40.

7.32. D.E. Smith, intro. to J.L. Heiberg, *Geometrical Solutions Derived from Mechanics: A Treatise by Archimedes*, trans. Lydia Robinson (Chicago, 1909), 1–2.

7.33. Heath, *Method, op. cit.* (1.3), 10.

7.34. Pappus, *op. cit.* (2.4), book 8.11, 2: 814.

7.35. *Ibid.*, 2: 841 and n. 4.

7.36. Hill, *op. cit.* (4.13).

7.37. Drachmann, *op. cit.* (5.34), 152–4. Oleson, *op. cit.* (5.4), 292, 294, 298.

7.38. A. Quacquarelli, 'La fortuna di Archimede nei retori e negle autori critiani antichi', Messina semin. matem. dell'Univ., *Rendiconti*, 1960–1, 5: 10–50, 16; Vitruvius, *op. cit.* (5.2), 199. Heath noted the mention of Archimedes only in the section on Archytas, also named by Vitruvius: T.L. Heath, *A History of Greek Mathematics*, two volumes, (New York, 1981), 1: 213.

7.39. Pappus, *op. cit.* (6.50), 7, 1: 37 and n. 81. (Aristophanes, *Clouds*, Tzetzes, *Aristophanes scholia*, line 1024, 621–2 (the power of mechanics); 39, Tzetzes, *Chil.*, 12, 965–71 (books, on the basis of which Heron, Anthemius and every other writer on mechanics wrote hydraulics, pneumatics, everything about the *baroulkos* and aquatic hodometry). Jones noted that this was an implausible list of (otherwise unattested) works of Archimedes.)

7.40. Sleeswyk, *op. cit.* (2.20) did not notice Tzetzes' claim that Archimedes wrote on hodometry. See 7.39.

7.41. Lewis, *op. cit.* (2.21).

7.42. Plutarch, *op. cit.* (7.4), 9: 121–3; Plutarch, *op. cit.* (1.4), 98–9. Plato's comment about Archytas may be sour grapes; Archytas was a successful philosopher-king. Despite Tannery's suggestion that this story was invented by Plutarch, Heath demonstrated that it matched Plato's opinions and was well founded. Heath, *op. cit.* (7.17), 1: 287 and n. 2, (P. Tannery, *La géometrie grecque*, 79, 80).

7.43. *Marc., op. cit.* (1.4), 14, 98.

7.44. Heath, *Method, op. cit.* (1.3), 10.

7.45. Santillana, *op. cit.* (7.17), 239–40.

7.46. Heath, *op. cit.* (7.38), 1: 295. (According to Heron, Plato knew of one of them in two forms. Heron, *Definitions*, 104, 66 Heiberg ed. Heath, *op. cit.* (1.3), xxxvi. According to Pappus (v. p. 352), Archimedes discovered 13 semi-regular solids.)

7.47. Heath, *op. cit.* (7.38), 1: 310–15, esp. 313 (distances as 1, 2, 3, 4, 8, 9, 27 in ascending order).

7.48. *Ibid.*, 313 (Macrobius, *In. somn. Scip.*, ii, 3.14).

7.49. *Ibid.*, 2: 18, n. 3 (Macrobius, *In. somn. Scip.*, ii, 3; cf. figures in Hippolytus, *Refut.*, 66.52 sq. ed. Duncker).

7.50. Chadwick, *op. cit.* (4.8), 82. Chadwick noted that 'The text of Hippolytus' *Elenchos* (iv, 8–10), severely corrupted in the manuscript tradition, gives figures for Archimedes' calculations of the distances between the planets, intended to illustrate Archimedes' contradiction of the Platonic dogma, and also cites a Ptolemaic attempt to reconcile them (iv, 11–12).' I have not been able to examine these references.

7.51. Heath, *op. cit.* (1.3), 222. Sloppy ideas about the sphere of the universe being linked to the dodecahedron do not match Archimedes' severe and precise approach to estimating the number of grains in the universe.

7.52. *The Republic of Plato*, 23.7, 524D–526C and 526C–527C, trans., intro., notes, F.M. Cornford (Oxford, 1941), 235–9.

7.53. *Marc., op. cit.* (1.4), 14, 99.

7.54. Georgiadou, *op. cit.* (6.10), 147–64, 147.

7.55. Drachmann, *op. cit.* (7.18), 2, 10, 11.

7.56. Georgiadou, *op. cit.* (6.10), 147–64, 147, 154, 162.

7.57. Cullham, *op. cit.* (6.10), 179–97.

7.58. *Ibid.*, 181, 182–3.

7.59. *Ibid.*, 184; 186–7. Polybius, *Rise, op. cit.* (1.10), 8. 3–7, 365–8. *Marc., op. cit.* (1.4), 17, 101; 14, 98.

7.60. Cullham, *op. cit.* (6.10), 186–7.

7.61. *Ibid.*, 179.

7.62. *Ibid.*, 188–9. *Marc., op. cit.* (1.4), 14.4, 98. She and Georgiadou fully agree. A novelist anticipated the central theme of their argument. According to Favre, in Suarès's novel *Hélène chez Archimède*, Helen of Troy personifies life in its most beautiful guise, while Archimedes personifies *l'esprit pur*. Y.-A. Favre, *La Récherche de la grandeur dans l'ouevre de Suarès* (Paris, 1978), 383. Suarès, *op. cit.* (1.2).

7.63. Cullham, *op. cit.* (6.10), 179–81.

7.64. *Ibid.*, 184.

7.65. *Ibid.*, 193.

7.66. Polybius, *Rise, op. cit.* (1.10), 8.7, 368.

7.67. *Marc., op. cit.* (1.4), 14, 99.

7.68. *Ibid.*, 99.

7.69. Cullham, *op. cit.* (6.10), 182. Sarton did believe that Plutarch's views were plausible. Sarton, *op. cit.* (2.2), 2: 70. Cullham, in claiming the contrary, missed this specific reference.

7.70. Proclus, *op. cit.* (6.52), 51.

7.71. *Marc., op. cit.* (1.4), 98–9.

7.72. Diodorus, *op. cit.* (1.8), 5.37.3, 3: 199.

7.73. They are not infrequently quoted, but the emphasis is almost always upon other matters.

7.74. Sarton, *op. cit.* (2.2), 2: 122–3 (Athenaios of Naucratis, 5.40–4).

7.75. Pappus, *op. cit.* (2.4), 8.1–2, 812.

7.76. Lloyd, *op. cit.* (7.17), 90–5. Georgiadou, *op. cit.* (6.10), 150–3 (Proclus), 153 (Pappus).

7.77. Being Cicero he solved it. I am indebted to George Engle for suggesting an equivalent in English to match that in American: 'a problem for a Philadelphia lawyer'. Sherlock Holmes said to Dr Watson in the *Red Headed League*, 'I see this is a three-pipe problem.' D.L. Simms, 'Problemata Archimedea', *T. & C.*, 1989, 30: 177–8.

7.78. Cicero, *Tusc. Disp.* (4.4) 1.2.5, 7. Plutarch mentioned nothing of Cicero's interest in Archimedes in his Life: Moles, *op. cit.* (7.20).

7.79. Silius Italicus, *op. cit.* (2.3), 14.337–55, 2: 297–9.

7.80. Livy, *op. cit.* (1.6), 24.34.3, 272.

7.81. Pliny, *op. cit.* (1.14), 7.37, 2: 589–91.

7.82. Firmicus, *op. cit.* (1.13), 210.

7.83. Clagett, *op. cit.* (6.89), 1: 2 and n. 3 (Cassiodorus praises Boethius for having restored in Latin the mechanician Archimedes to the Sicilians; see Cassiodorus, *Liber variorum*, 1.Ep.45, 46, *op. cit.* (4.8), 168– 70 (order for water-clocks and record of delivery)).

7.84. Gibbon, *op. cit.* (4.8), 4: 140–1. Gibbon is alone (of all the many historians and commentators who have noted the extraordinary polymathy of Boethius) in commenting on his practical abilities.

7.85. Chadwick, *op. cit.* (4.8), 23.

7.86. Pappus, *op. cit.* (6.52), 1: 37 and n. 81. (Aristophanes, *Clouds*, Tzetzes, *Aristophanes scholia*, line 1024, 621–2 (the power of mechanics); 39, Tzetzes, *Chil.*, 12, 965–71).

7.87. *Ibid.*, 1: 39 (Tzetzes, *Chil.*, 11, 586–641 (on geometry and optics, particularly lines 586–610, 616–18)), 40, 41, n. 85.

7.88. Simms, *op. cit.* (7.78), 177–8.

7.89. Tzetzes knew of Plutarch. It is uncertain whether he knew the passages in *Marcellus* since he referred only to Diodorus and Dio Cassius in commenting upon the role Archimedes played in the siege of Syracuse.

7.90. *Marc., op. cit.* (1.4), 14.5, 98–9.

7.91. Vitruvius, *op. cit.* (5.2), 199; Marsden, *op. cit.* (6.9); Lloyd *op. cit.* (7.33), 95–6.

7.92. Knorr, *op. cit.* (2.37), 210–13, 210–11, 213.

7.93. Vitruvius, *op. cit.* (5.2), work as a whole and book VII, introduction, para. 14, 199.

7.94. Green, *op. cit.* (6.5), 643. Green noted how antipathetical this was to the spirit of Aeschylus.

7.95. Heath, *op. cit.* (1.3), 260. *Carmen de Ponderibus et Mensuris.*

7.96. Finley, *op. cit.* (6.98), 176–98, notes 273–4, addendum 274–5, 176, 181 (Archimedes' inventions purely military). Lloyd, *op. cit.* (2.36), 321–52, 337. Green, *op. cit.* (6.5), 367, esp. ch. 27, 467–79, 471, 476–7. Green occasionally overstretched his case: Archimedes not based in Alexandria, Hieronymus was Hieron's grandson, not son (*ibid.*, 464). More seriously, he believed that 'It is probably, in the last resort, better to drop with fatigue from hoeing potatoes than to be poisoned slowly, by synthetic food additives and carbon monoxide' (*ibid.*, 234). Professor Green must have misunderstood his sources. Carbon monoxide can be quite quick. Did he mean carbon dioxide and global warming? I for one will take my chance with synthetic food additives against hoeing potatoes (sometime Chairman, Advisory Committee on Research and Development into Environmental Protection, Commission of the European Communities). For more errors on Archimedes, see section on Arts of War.

7.97. Finley, *op. cit.* (6.98), 189 (Tiberius), 192 (Vespasian). It is difficult to believe either tale. No one has ever made unbreakable glass. The Romans did use labour-enhancing devices; one is to be found on Trajan's Column. These discrepancies reinforce the notion of hostility.

7.98. Green, *op. cit.* (6.5), 643.

7.99. Finley, *op. cit.* (6.98), 185, 177, 179.

7.100. *Ibid.*, 181–3.

7.101. P.-M. Duval, 'The Roman Contribution to Technology', in *A History of Technological Invention: Progress through the Ages: The Origins of Technological Civilization*, ed. M. Daumas, trans. E.B. Hennessey (London, 1969), 1: 216–59, 257–8.

7.102. Finley, *op. cit.* (6.98), 179, 185. A Master of a Cambridge College criticizing expenditure on amenities is an unexpected and charming paradox.

7.103. J. Deshayes, 'Greek Technology', in *op. cit.* (7.101), 1: 181–215, 181 (except for war and temples). Lloyd, *op. cit.* (7.17), 95–112.

7.104. Green, *op. cit.* (7.17), 239.

7.105. Finley, *op. cit.* (6.98), 185.

7.106. Oleson, *op. cit.* (5.4), 406–8.

7.107. Lloyd, *op. cit.* (2.36), 351. Forbes, *op. cit.* (5.7), 604.

7.108. Green, *op. cit.* (6.5), 467.

7.109. It matched that in religion; cf. Gilbert Murray, 'The Failure of Nerve', in *Five Stages of Greek Religion* (London, 1946), xiii (attributes the expression to J.B. Bury).

7.110. Finley, *op. cit.* (6.98), 176 (not very much); Green, *op. cit.* (6.65), 467.

7.111. Duval, in *op. cit.* (7.104), 1: 216–59.

7.112. Marsden, *op. cit.* (6.10), 48.

7.113. White, *op. cit.* (5.20), 192.

7.114. *Ibid.*, 116. J. Harvey-Jones, *Getting It Together* (London, 1991), 338. 'Technologically proud and able companies always want to make big inventions. Big leaps forward are much more satisfying than small incremental changes. Yet, making money from a great invention is notoriously slow and difficult. It is the small innovations, targeted directly to someone's needs, that produce the quick and generous payback ... They derive from the hot breath of the customer on your neck, combined with the technical mastery of your own field.'

7.115. White, *op. cit.* (5.20), 27–48 (main areas), 58–172 (seven technologies), 189–220 (appendices), 221–40 (tables).

7.116. *Ibid.* (5.20), 61–2 (grain-harvester), 30– 2 (agriculture), 32–4 (water-raising), 34, 35 (mining), 36, 37, 123–34, 125, 126 (metallurgy), 37, 38, 39 (pottery), 39 (textiles), 42 (glass), 43 (mosaic), 44, 45, (heating), 79, 81 (building), 83–5, (building technology), 85 (kiln-face brick), 86–9 (concrete), 93 (roads), 97–9 (bridges), 101–4 (dams), 99–101 (cuttings), 140 (land transport), 141–56 (sea), 157–72 (hydraulic).

7.117. Deshayes, *Daumas, op. cit.* (7.101), 213.

7.118. R. Fiengo, Letter, in *Scientific American*, August 1984: 5.

7.119. Wilson, *op. cit.* (6.40), 757.

7.120. S. Applebaum, 'Animal Husbandry', in Wacher, *op. cit.* (5.4), 2: 504–20, 519.

7.121. White, *op. cit.* (5.20), 179.

7.122. Oleson, *op. cit.* (5.4), 406–8.

7.123. J.B. Ward-Perkins, *Roman Imperial Architecture*, The Pelican History of Art (Harmondsworth, 1981), 97–120, 111, 121. D.R. Hill, *A History of Engineering in Classical and Medieval Times* (London and Sydney, 1984), 52 (greatest civil engineers, antiquity, sheer size, scope, their activities).

7.124. Hall, 'Military Technology', in Singer *et al., op. cit.* (5.7), 695–730, 713 (Ammianus Marcellinus, *The Histories*, 23.4.4–8, trans. J.C. Rolfe (London, 1937), 2: 326ff.).

7.125. O. Wikander, 'Water-power and Technical Progress in Classical Antiquity', *op. cit.* (2.20). D.L. Simms, 'Water-driven Saws, Ausonius and the Authenticity of the *Mosella'*, *T. & C.*, 1983, 24: 635–43.

7.126. A. Roland, 'Secrecy, Technology and War: Greek Fire and the Defence of Byzantium, 678–1204', *T. & C.* 1992, 33: 655–79.

7.127. Wikander, *op. cit.* (2.20), 81–2.

7.128. *Ibid.*, 68–84, 81.

7.129. Field and Wright, *op. cit.* (4.18), 87 and 136. They rightly emphasized that the form of the device, the material from which it was made and its detailing suggested that it was a common enough object; once again an artefact had revealed the inadequacy of literary evidence.

7.130. C.E. Conophagos, 'The Ancient Athenian Silver Mines', in *op. cit.* (2.20), 121–30. Gille, in Singer *et al., op. cit.* (5.7), 2: 639 (partial mechanization continued, notably in flour milling and olive oil); 2: 630. Deshayes, *Daumas, op. cit.* (7.101), 190–1.

7.131. *Ibid.*, 199 (new styles of pottery; larger kilns), 205 (assembly line production).

7.132. Gille, *op. cit.* (5.7), 2: 638.

7.133. Deshayes, *Daumas, op. cit.* (7.101), 213.

7.134. J.B. Ward-Perkins, *Roman Imperial Architecture*, The Pelican History of Art (Harmondsworth, 1981), 97–120, 111, 121.

7.135. Wilson, *op. cit.* (6.40), 2: 361–400.

7.136. J.E. Stambaugh, *The Ancient Roman City* (Baltimore, 1988), 156; T.W. Potter, *Roman Italy* (London, 1987), 52, 80.

7.137. Deshayes, *Daumas, op. cit.* (7.101), 199 (new styles; larger kilns), 205 (assembly line).

7.138. Duval, *Daumas, op. cit.* (7.101), 231, 233. When did Wedgwood reach that output? Comments about the absence of craft revolutions provokes the question 'when does a craft become an industry?' If the definition is the transformation of the Industrial Revolution, then it occurred nowhere before the eighteenth century. It excludes a great deal.

7.139. A. Trevor Hodge, 'A Roman Factory', *Scientific American* 1990, 263: 58–64, 64 (slavery ceased to be a source of labour after *c.* AD 100).

7.140. Deshayes, *Daumas, op. cit.* (7.101), 206.

7.141. Oleson, *op. cit.* (5.4), 406–8.

7.142. Wikander, *op. cit.* (2.20), 74–5.

7.143. M.J.T. Lewis, 'The Origins of the Wheelbarrow', *T. & C.*, 1994, 35: 453–75, 475 and 456.

7.144. Green, *op. cit.* (6.5), 474 and 856–7, n. 71.

7.145. J. Needham, 'The Prenatal History of the Steam-engine', in *Clerks and Craftsmen in China and the West* (Cambridge, 1970), 136–202. A door in the library at a Rothschild Mansion, at Ascott in Buckinghamshire, has false leather bindings of lost and imaginary books: *Archimedes on the Steam Engine*, two volumes.

7.146. M. Scott, 'A New Theory of Endogenous Growth', *Oxford Review of Economic Policy*, 1993, 8: 29–42.

7.147. The organization of the marble trade is another example of large-scale trade dominated by the state.

7.148. Sextus Julius Frontinus, *The Stratagems and the Aqueducts of Rome*, trans. C.E. Bennett (London, 1925).

7.149. When the Office of the Minister for Science was created in 1959, a sardonic scientific civil servant was recorded as commenting that the wall facing the minister's office in Richmond Terrace, Whitehall, was lit by gas lamps. See Rose and Rose, *op. cit.* (1.9), 84. After more than 35 years, I abandon my disguise, but not my disgust.

7.150. D.J. De Solla Price, 'Is Technology Historically Independent of Science?', *T. & C.*, 1965, 6: 553–68. P.M.S. Blackett, PPRS, was its most distinguished expositor in the 1960s (personal communication). It was a much fancied theory, but it had no foundation in fact, least of all at the time it was most fashionable among those *fanning* the white heat of technology. The process of innovation, then as now, was much more complex.

7.151. Marsden, *op. cit.* (6.10), 8. Diodorus, *op. cit.* (1.8), 14.50.4, 6: 153.

7.152. Oleson, *op. cit.* (5.4), 406–8. Deshayes, *Daumas, op. cit.* (7.104), 213.

7.153. Price, *op. cit.* (5.19), 51. Field and Wright, *op. cit.* (4.18).

7.154. Gibbon, *op. cit.* (4.8), 2: 80–1.

7.155. Gille, *op. cit.* (5.7), 2: 638.

7.156. *Marc.*, *op. cit.* (1.4), 14, 98 (Archimedes had done this in the past because Hieron persuaded him).

7.157. Gille, *op. cit.* (5.7), 2: 633.

7.158. *Ibid.*, 2: 638. Unfortunately, no references to these documents.

7.159. White, *op. cit.* (5.20), 13.

7.160. Vitruvius, *op. cit.* (5.2), preface, 1, 3–4.

7.161. Ward-Perkins, *op. cit.* (7.134), 61, 73, 123, 164, 216, 263, 264; R.P.H. Green, *The Works of Ausonius* (Oxford, 1991), 124, line 303; Vitruvius, *op. cit.* (5.2), 199, 313, 315, 316, 317; Sarton, *op. cit.* (2.2), 1: 363.

7.162. Gibbon, *op. cit.* (4.8), 4: 184 and notes. The cult of personality was earlier than Stalin.

7.163. M. Kline, *Mathematics in Western Culture* (New York, 1964), 61–2. Kline overestimated the practical value of some of their achievements, such as steam power used to drive automobiles, and believed in those burning mirrors. See section on 'Art of War'. Hacker, *op. cit.* (6.69), 34– 50.

7.164. G.G. Joseph, *The Crest of the Peacock: non-European Roots of Mathematics* (Harmondsworth, 1991), 8; 56, note (Plutarch cited for his consequences upon Greek thought).

7.165. W.W. Tarn, 'Macedon and the East', in *The Root of Europe*, ed. M. Huxley (London, 1952), 17; F.W. Walbank, *The Hellenistic World* (Glasgow, 1981), 64 (close ranks, Greekness), 65 (closed circle), 113 (Greek and Macedon ruling class) (Polybius, *Hist.*, 34.14, 1–5, shows Greek stock had not forgotten Greek customs); S. Price, 'The History of the Hellenistic Period', in *Greece and the Hellenistic World, op. cit.* (6.40), 309–31, 316 (generally, . . . only way to advance in society, . . . adopt Greek culture).

7.166. R. Jenkins, *Asquith* (London, 1964). In that brilliant biography of Asquith, the Liberal Prime Minister from 1906 to 1916, the only mention of British industry is the strikes of 1911–13. That discussion is entirely confined to their political and social causes and consequences. There is nothing on the underlying reason, Britain's loss of competitiveness, of which Asquith, like the rest of his Cabinet, was almost certainly unaware. Not that the Opposition was any better; those of its members who supported the doctrine of Empire Preference were unconsciously accepting that Britain's goods were incapable of competing in an open market with the technologically advanced German and American economies.

7.167. Duval, *Daumas, op. cit.* (7.101), 1: 258.

7.168. Lynn White, *Medieval Technology and Social Change* (Oxford, 1964). Lynn White, 'The Expansion of Technology, 500–1500', in *The Fontana Economic History of Europe: The Middle Ages,* ed. C. Cipolla (Glasgow, 1972), 143–71.

7.169. Oleson, *op. cit.* (5.4), 406–8.

7.170. Hodge, *op. cit.* (7.126), 58–64, 61–2.

7.171. Wikander, *op. cit.* (2.20), 68–73. Simms, *op. cit.* (7.127).

7.172. J. Gimpel, *The Medieval Machine: The Industrial Revolution of the Middle Ages,* 2nd edn (London, 1993), 9. This is not unreasonable. Monks were the first intelligent people to go back to the land and it is hard to believe that the third and later generations would not seek ways of lightening their burdens. However, the Ptolemies exploited water-lifting devices on a comparable scale. Oleson, *op. cit.* (5.4), 406–8.

7.173. B. Gille, 'The Medieval Age of the West: Fifth Century to 1350', in *Daumas, op. cit.* (7.101), 1: 422– 571, 455.

7.174. Wikander, *op. cit.* (2.20), 81 (archaeology), 79 (heavy plough).

7.175. White, *op. cit.* (7.17), 211–20, 213 (Landels), 217 (Ktesibios), 236–7 (slavery).

7.176. Duval, *Daumas, op. cit.* (7.101), 219. The distinction remains. The National Trust found this out the hard way in not using Roman cement to repair Hadrian's Wall.

7.177. D.L. Simms, 'Was There a Technological Decline and Fall of the Western Roman Empire?' (in preparation.)

7.178. White, *op. cit.* (7.17), 223.

7.179. J.D. Hughes, *Pan's Travail: Environmental Problems of the Ancient Greeks and Romans* (Baltimore, 1994), esp. 112–29. Gimpel, *op. cit.* (7.177), 75–8.

7.180. G. Hollister-Short, 'The Other Side of the Coin: Wood Transport Systems in Pre-industrial Europe', *Hist. Tech.*, 1994, 72–97.

7.181. Duval, *Daumas, op. cit.* (7.101), 255.

7.182. If some of that state's expenditure was wasteful, in the sense of not providing economic returns, then so was some private expenditure on windmills in the Middle Ages. J. Langdon, 'The Birth and Demise of a Medieval Windmill', *Hist. Tech.*, 1992, 14: 54–76, 54–5 (built partly to bolster feudal authority, derelict in 20 years). Did Finley never consider the railway mania in nineteenth-century Britain and the even less excusable public investment in nuclear energy this century?

7.183. White, *op. cit.* (5.20); Hassan and Hill, *op. cit.* (5.19); Lynn White, *op. cit.*, 1964 (7.168); Lynn White, *op. cit.*, 1972 (7.168); Gimpel, *op. cit.* (7.172); J. Needham, *Science and Civilisation in China* (Cambridge, 1954).

7.184. Vitruvius, *op. cit.* (5.2), 10, 281–319; White, *op. cit.* (5.23), 13 (Pliny, *Nat. Hist.*, 2, 117–18).

7.185. Hall, in Singer *et al., op. cit.* (5.7), 695–730, p. 719.

8.1. Andrew Undershaft, 'Major Barbara', in *The Complete Plays of Bernard Shaw* (London, 1936), 497. Petrarch may have been correct in assuming that Archimedes prepared his weapons for the defence of his country's liberty, or those that survived as a client state of Rome. Having myself worked for someone who at that time was also Chief Scientific Adviser to the Ministry of Defence, Plutarch's tale of a reluctant mathematician dragged away from his books by his kinsman, Hieron, does not ring true.

8.2. Knorr, *op. cit.* (2.6), 441.

8.3. Until Kepler adapted the technique to estimate the size of wine casks, some 1880 years later, no one had found any practical use for it. J. Kepler, *Stereometria doliorum vinariorum,* quoted from M. Caspar, *J. Kepler,* trans. C. Doris Hellman (London, 1957), 233–4.

Moriscos and Marranos as Agents of Technological Diffusion

THOMAS F. GLICK

I

The propensity of ethnic minorities to specialize in distinctive techniques and thus to play a role in their diffusion is well known to cultural anthropologists. It is, indeed, a commonplace that in ethnically stratified societies, ethnic groups tend to monopolize certain craft specialities. The related subject of emigrant groups as agents of diffusion in their new countries of techniques brought from their original home is part of the same discussion. Cultural contact always implies the existence of some technological disparity between one group and another and therefore a kind of technological determinism is presumed, 'as if imbalances in the larger set of cultural systems can result in a flow of information from one culture to another according to general economic principles'.[1]

The history of Spanish Muslims and Jews illustrates both processes: techniques practised within ethnic enclaves before their expulsion and, afterwards, emigrants as possessors of new techniques not before known in their new countries. But inasmuch as neither topic has been studied in any depth, we face a series of parallel, even symmetrical, myths referring to the technical skills of both groups.

Elsewhere, I have commented on Américo Castro's peculiar assertion with respect to ethnic enclaves. According to his conception, the medieval communities of peninsular Jews and Muslims practised techniques not used by or known to Christians who, in this respect, colonized their own minorities: 'The Christians both then and later lived behind the barricade of their sense of the "intrinsic value" of their land and their persons, and thus the inner posture of the Christian caste turns out to have the same form in the tenth century and in the nineteenth: in the earlier period, adoptions from the Muslims; in the later, European importations.' He goes on to say that Christians continued to seek innovations and ideas from Jews and Muslims alike until their respective expulsions.[2] The paradox is attractive, but inexact. Castro is right up to a point: in the tenth century there was certainly a technical differential between Christians and Muslims, which stimulated the flow of techniques from the latter to the former. Later, upon ingesting masses of Jews and Muslims, especially after the conquest of Toledo in 1085, Christians could, for a while, take advantage of the skills of the two ethnic enclaves. But, with the passage of time, those same techniques were adapted by Christians and the technical archive of the three religious groups tended towards homogeneity. Castro's error, with regard to the lower Middle Ages and, in the case of Moriscos, the sixteenth century, was to have confused an ethnic division of labour (a social and economic phenomenon) with a technological division of labour, which has quite different implications. If the distinction is not made, the historian runs the risk of falling into a kind of popular diffusionism, attributing certain techniques or products to Muslims or Jews simply on the basis of their exterior appearance or design or as the result of some imagined process.

All the received views concerning the technical prowess of the expelled groups are currently being revised. With respect to the Moriscos, for example, assertions of Castro and Julius Klein regarding their supposed monopoly of agriculture in certain regions of Spain have been challenged,[3] just when a number of Tunisian authors have attacked – with perfect symmetry – older views regarding the Morisco contribution to North African agriculture. Or, in the case of Marranos, recent authors have attacked their technological contribution to the economies of both Holland and the Ottoman Empire.

Inasmuch as the Moriscos and many Marranos settled in the Turkish Empire, the myth of Spain's technological loss has an interesting counterpoint in the complementary myth of the Islamic world's gain in accepting them. The myth no doubt arose among both groups and was continued by their historians and propagandists ever since.

In the present essay I cannot hope to resolve these questions, especially in view of the lack of research on many of the specific techniques discussed. My objective is the modest one of commenting on some of this literature in order to point out promising avenues for future research.

II

The Morisco contribution to agriculture in the countries of North Africa is the central theme of Morisco technological importation, especially in view of the traditions associating the Andalusis with great dexterity in agrarian techniques, especially hydraulic ones. They were also presumed to have been the inheritors of the Andalusi agronomical tradition of Ibn Baṣṣāl, Ibn al-ᶜAwwām, Ibn Wāfid, Ibn Luyūn and the other agronomists of Seville, Toledo and Granada. Insofar as written materials are concerned, the Morisco agronomical tradition has hardly been studied. A sixteenth-century Moroccan from an Andalusi family named Muḥammad ibn ᶜAlī al-Shaṭībī al-Andalusī wrote an 'Epistle on the Vocation of Agriculture' (*Risāla san'at al-filāḥa*), a compilation of Andalusi agronomical literature in 40 chapters.[4]

It is more than likely that the Moriscos introduced American food crops and medicinal plants in North Africa. With respect to *materia medica*, Razouk includes in his list of Hispanisms introduced into colloquial Moroccan speech the word *taba*, tobacco,[5] a plant originally received as a medicine. It was due to Moriscos that the purgative *mechoacan* was introduced into France, as well as many other Mexican simples like *ocozotl* (liquidambar).[6] We can presume that a similar process took place in the Magrib, with respect not only to medicine but to new crops as well. The introduction by Moriscos of maize, tomatoes, certain varieties of beans, American pepper (chili) and opuntia is documented.[7] The Moroccan physician al-Ghassānī (1548–1611) published an Arabic text that Vernet believes to be either an adaptation or a translation of the *Historia natural* (1574) of Nicolás Monardes, a key work in the description of the new American medicinal flora, including tobacco.[8] According to a recent author, the introduction by Moriscos of the American cultivars constituted a second agricultural revolution in the Magrib.[9]

The received view of Morisco impact on North African agriculture is that of J. D. Latham.[10] This author presents a series of observations by European travellers, mainly of the eighteenth century, all of whom commented on the agricultural activities of the descendants of Moriscos, in particular the cultivation of olive (especially in the Cape Bon peninsula of Tunisia) and fruit trees and the modernization of techniques of olive, grapevine and cereal cultivation. Moriscos are also credited with the extension of irrigation systems and the application of irrigation to olive trees.

Lately, several Moroccan writers have raised doubts concerning the significance of the Morisco contribution. Thus, for Kassab, irrigated fruit tree cultivation in Testour was nothing more than a small island in the midst of a pastoral sea modified by extensive open-field cereal cultivation over which the Moriscos had no influence at all.[11] For El Aouani, the Moriscos only improved and extended pre-existing irrigation systems,[12] while for Sethom not only irrigated agriculture but also arboriculture in the region around Cape Bon was the work of 'indigenous populations', not of Moriscos.[13] Nevertheless, the three articles just cited are of a general nature and lack the linguistic, technological and historical evidence that would contribute to the resolution of the questions raised. One would have to establish, for example, whether the Moriscos irrigated wheat in Tunisia in order to increase yields, as they had done previously in Aragón.[14]

Razouk gives a list of words that have entered Moroccan colloquial usage owing to the immigration of Moriscos.[15] One notes immediately that a whole series of terms related to the wheeled cart are represented in Moroccan dialectal Arabic by Hispanisms: *kārū* (Spanish, *carro*), *kūtshī* (Spanish, *coche*) – both are *ʿaraba* in classical Arabic – and *ruayda* (Spanish, *rueda*; *ʿajala* in classical Arabic). Here we confront a technological question of the first order. We know that the two-wheeled agricultural cart practically disappeared from the medieval Arab world. Richard Bulliet explains the phenomenon as the result of the interaction of a number of factors: first, a saving of 20 per cent in transport by camels over the cost of transport by cart; second, a reciprocal relationship between the decline of Roman roads and the introduction of the camel; third, the significant role of the North Arabian camel saddle in affecting the transition and consolidating the domination of nomadic styles over those of sedentary farmers.[16]

Nevertheless, Bulliet notes that the cart did not disappear completely, but survived in a few regions, especially Tunisia. The use of carts by the Aghlabid emirs of Tunis in the eleventh century is documented, as is their use in Muslim Sicily, conquered by the same Aghlabids. A camel-drawn cart used in Morocco was known as the Sicilian *ʿaraba.*[17] Bulliet observes that in coastal zones of Tunisia the cart was known by the classical term *ʿaraba,* while in the southern oases it was called by the Latinism, *kirrīta* – that is, Italian and Spanish *carretta, carreta,* a form documented in the *Dictionary* of Pedro de Alcalá as *carréta,* pl. *carárit.*[18] Bulliet associates the survival of the cart in Tunisia with the existence there of a type of camel harness which is related to harnesses, and described with similar terms, in Spain, Italy and Portugal, which, for him, constitutes a technically distinct zone, closely related to Tunisian agrarian techniques, while the south of France belongs to a different technological and cultural zone. One of the terms in question is that for the straps placed over the shoulders of the animal to connect the cart shafts: *ghunj* in Tunisian Arabic, *enganche* in Spanish, *gancio* in Italian.[19]

Bulliet does not mention the Moriscos. But the introduction of the agricultural cart with two wheels, especially in Morocco, but also in Tunisia, has been attributed to them by others. Thus Latham asserts that Moriscos introduced the *kirrīta* to north-east Tunisia.[20] Likewise, Bulliet's hitching strap – *enganche* – is known in Morocco by the Hispanism *ghānjū,*[21] easily related to the Tunisian *ghunj.* That is, for Bulliet the Tunisian *kirrīta* is a Roman survival and its associated rig diffused from the Islamic world to Europe, through Sicily and Spain. For Latham and other Morisco scholars both the term and the normal use of the cart were a Morisco introduction of the seventeenth century.

Moriscos were prominent in the military life of the Magrib in the sixteenth and seventeenth centuries. Muḥammad Zarqūn al-Kahiya, of Guadix, a veteran of the second Alpujarras war, organized the artillery corps for ʿAbd al-Malik, pretender to the Moroccan throne, half of whose army was made up of Moriscos, especially his musket corps.[22]

It was a precept of Ottoman law with respect to holy war (*jihād*) that the infidel must be opposed 'with his own weapons and devices'.[23] Such an ideology makes intelligible a policy of appropriation of foreign military technology. In this respect, an important contribution to *jihād* made by a Morisco was the *Manual of Artillery* of Ibrāhīm ibn Aḥmad ibn Ghānim al-Andalusī, an artilleryman who arrived in Tunis from Spain in 1609 or 1610.[24] Ibn Ghānim had been a sailor in the Spanish navy and served in the West Indies, where he acquired skills in firearms. While in the service of Yūsuf Dey of Tunis he decided to write an artillery manual, similar to Spanish texts of his day, for the use of Tunisian artillerymen whom he deemed technically deficient. He began to write the book in 1630 in Spanish and later found a translator who produced the book known as

*Kitāb al-izz wa'l-manāfī' lil-mujāhidīn fi sabīl illāh bi'l-madāfi*ᶜ (*The Book in Which One Seeks Triumph and Advantage when Fighting against the Infidel with Military Stores*). Ibn Ghānim seems to have depended above all on the *Plática manual de artillería* of Luis Collado (first Spanish edition, 1592), inasmuch as most of his 50 chapters are either translations or else summaries of passages from Collado.[25] The translator, the Morisco Bejarano, appears to have also translated into Arabic the *Aritmética algebráica* (Valencia, 1522) of Marco Aurel Alemán.[26]

In spite of some notices about the participation of Moriscos in the silk industry[27] and in pottery, the only craft industry studied in any depth is the fabrication of *shāshiya*, the typical Tunisian cap, which was a kind of Morisco monopoly.[28] The fulling vocabulary is filled with Hispanisms; for example, the fulling mill itself, *bīlāḍa* (Spanish, *pilada*), and mallet, *māṣu* (Spanish, *mazo*). After fulling, the woollen cloth was said to be *shāshiya kruḍu*, that is, 'raw' cloth (Spanish *crudo*). All three steps of the carding process were known by Hispanisms: *būṭār* (Spanish *botar*, throwing), *ābācār* finishing, *acabar*, by metathesis) and *āfīnār* (refine).[29] The parts of the carding machine, which consists of five cylindrical cards, were also described by Hispanisms, from the *bī* (foot; Spanish, *pie*) of the carder to its *kbīsa* (head; Spanish, *cabeza*). The same was true of dyeing and packaging terms.

Moriscos also introduced the term *mākīna* (machine; Spanish, *máquina*) in substitution for the classical Arabic, *alat*.[30] In the medieval discussion of machines the Arabic equivalent of the Greek term was *ḥīla*, pl. *ḥiyal*, referring to 'any device that allows one to overcome the natural resistance and thus performs actions contrary to the natural tendency'.[31] In this sense, each *ḥīla* gives form to a specific physical idea. According to al-Fārābī, *ḥiyal* were used to give palpable demonstration to the various processes that natural bodies undergo, making them perceptible to the senses.[32] It is well known that the Castilian word *máquina* is an Italianism which entered common usage at the end of the sixteenth and beginning of the seventeenth centuries, when it appears in the works of Ambrosio de Morales or Cervantes in the sense of *ingenio*, the word that it replaced.[33] It is interesting to note, with respect to Moriscos and carts, that in addition to the meaning of an instrument or machine in general, the word *máquina* began to be used in the seventeenth century in the sense of vehicle.[34] It may be that the Moriscos introduced the term in the Magrib with this connotation.

III

With regard to the technological achievements of the Marranos, I shall limit my remarks to their actions in the Islamic world. Here, the discussion has centred in great part on the textile industry, especially that of Salonica – in Ottoman Greece – well known since the sixteenth century as a Spanish-speaking city, so great was the number of Sephardic exiles who settled there. In Greece, the Spanish Jews introduced a new and modern woollen textile industry, that is the production of *cuha* or *cuka*, the equivalent of Spanish *velarde*, broadcloth. More heavily felted than the common cloth of the region, the new cloth was the result, according to Braude, of 'a particular mechanical technology which the Jews brought with them from the West'. This technology produced a stronger, cheaper cloth than that which it replaced, owing to the use of the water-driven fulling mill (see Section IV below).[35] In the immediate environs of Salonica there were some three hundred persons employed in fulling mills, whose use then diffused to other areas of Greece.

There is an interesting description in Hebrew of this process in a sixteenth-century *responsum*, or rabbinical opinion, emitted by Rabbi Samuel de Madina of Salonica:

> I will preface with an introduction to cloth manufacture in Salonica. After the cloth has been woven it is given to the fuller [*batañero*] who fulls it. Then it is given to the comber [*perharo*][36] to perform the task which is called combing [*bettildar*].[37] Then it is given to shearing [*tundir*]. After this task is done they see if the cloth is pure and clean of oil. If it is not as clean as it should be it is given to another man to degrease it and this task is called bleaching [*desteñir*]. Nitre and water are placed upon the cloth. After it has been degreased the comber takes it to comb [*perhar*] it a second time. Afterwards the tenter-man [*tendirdor*][38] takes it to tenter it in the sun of the courtyard as is the custom.[39]

Braude concludes, in replying to authors who have raised doubts concerning the Spanish origin of the woollen industry of Salonica, that Spanish terminology suggests a Spanish origin, even if it does not prove the case definitively. Spanish Jews also mounted a similar textile industry in the Palestinian city of Safed.[40]

There are various general references to the role of Spanish Jews in the transfer of technology to the Turkish world. Two French travellers in Istanbul in the mid-sixteenth century noted that Marranos 'were the men who have taught the Turks how to trade and to deal with those things that we use mechanically'. It was also said that the political figure Giovanni Soranzo prevented the expulsion of the Jews from Venice in 1571, asserting that it had been 'Jews expelled by the kings of Spain' who had given the Turks their power in the form of artisans who knew how to make cannons, cannonballs and other arms.[41] In the same sense, we can cite the famous passage in the *Viaje de Turquia* by the pseudonymous Cristóbol de Villalón, where he explains that the Turkish artillery 'had no masters to teach them (particularly how to mount pieces on carriages) until the Jews were thrown out of Spain. They showed them how, as well as how to fire muskets, to make forts and trenches, and whatever devices and strategies there are in war, because before they [the Turks] were no better than animals.'[42]

Still, solid evidence of the participation of the Sephardim in new industries is not so easy to come by. Printing constitutes the main exception: David Nahmias, a Spanish Jew, founded a Hebrew press in Istanbul, perhaps as early as 1503. He was followed by the Gedaliahs, a

Jewish family from Portugal, in 1512, and afterwards a number of presses were established in Salonica. In Morocco, Samuel Isaac printed Hebrew books in Fez between 1516 and 1524, and had learned the art in Lisbon, whence he imported his machinery.[43]

The processing of tobacco was, from the sixteenth century on, one of the principal industries of the Jews of Salonica,[44] which is suggestive of contacts between the exiles and *conversos* in Spain. In Savoy too, a Marrano named Jacobo Moreno played a pre-eminent role in this industry and grew, as well as processed, tobacco. Moreno held a kind of monopoly over the commercialization of tobacco in Savoy. Apart from him, only those who cultivated it for medicinal uses were given permission to sow it.[45] In 1535, Juan Robles, a *converso* glassblower from Cadalso de los Vidrios (Toledo) was condemned to death *in absentia* by the Inquisition. He had already emigrated to Fez, where he lived as a Jew, practising his art with tools possibly sent from Toledo by his mother. He declared his intention to move on to Jerusalem, but whether he did so is not known.[46] The story is interesting, not only because it is preserved in Inquisitorial records, but because it illustrates a process of the transfer by *conversos* of Spanish techniques to the Islamic world.

Techniques of applied astronomy also owe something to the Spanish Jews. When the famous astronomer Abraham Zacuto left Portugal around 1497, he resided for a while in Tunis where, we might suppose, he circulated his *Almanach perpetuum*, inasmuch as an Arabic translation by one Yusuf al-Andalusi (Joseph the Spaniard) appeared there.[47] Meanwhile, in Morocco, Zacuto's astronomical tables were translated into Arabic by the Morisco Bejarano, whose Arabic name was Ahmad b. Qāsim al-Hajarī.[48] It is interesting to note there the collaboration between a Jew and a Muslim, both expelled from Spain. In 1505 Zacuto moved on to Egypt, where his tables were widely disseminated in the sixteenth century.[49]

A form of practical astronomy was also diffused in Ladino, the Spanish of the Jewish exiles written in Hebrew characters. Moshe Almosnino, a rabbi born in 1510 in Salonica, where he served the community of Catalan Jews, wrote a treatise on the construction and use of the astrolabe (1560). This treatise appears to have been faithful to the Ptolemaic tradition of Jewish astronomical writers from Abraham ibn Ezra on, all of whom wrote about this practical instrument which the devout used to calculate the religious calendar. It is curious that Almosnino was so practised in astronomy that he wrote his treatise on the 'astrolabe ... in Castilian Romance', from memory, as he himself recounts, being away from home and 'without books'.[50]

That Spanish Jews modernized medicine in the Turkish Empire had already become a cliché by the eighteenth century, when Voltaire praised a Jewish physician, Daniel de Fonseca, who had been adviser to the sultan at the beginning of the century, as 'perhaps the only philosopher of his nation'.[51] But the contents of medical knowledge that the Jews carried with them to the East have yet to be studied.

A Ladino translation of Luis Lobera de Avila's *regimen sanitatis*, titled *Vergel de sanidad* (1530, 1542), suggests that Spanish physicians in Turkey remained in contact with Iberian medicine. The most useful part of the *Vergel* was a chapter on plague that later on was generally published separately. The edition of 1542 also contains some norms for those who travel by land and sea, an attractive reading for the large Spanish-speaking Jewish merchant community.[52]

In the Ottoman court itself there was an important and interesting figure, born in Granada, named Moshe Hamón, a physician in the service of Suleyman the Magnificent and who appears in Turkish documents as Hamon-ugli. To Hamón is owing the first treatise on dentistry written in Turkish and perhaps another little treatise on pharmaceutical compounds signed Musa Calinus (Galen) el-Israeli. This treatise, whether Hamón wrote it or not, is interesting for its eminently medieval contents, with citations of Averroes and Arnau de Vilanova. It seems clear that at least the first generations of Sephardic physicians did not introduce the new concepts of European Renaissance medicine in the Turkish Empire.[53]

In the *Viaje a Turquia*, the author ridicules the Jewish physicians of the Ottoman court, who practised 'almost by inheritance. Their fathers left them their thinking cap and a book which tells in Romance to use such and such a remedy to cure such and such a disease'[54] – that is, exactly the kind of work typified by the *Vergel de sanidad*, a little book that circulated in Ladino, as we have seen. Such texts were quite widely disseminated. Three Ladino fragments on medical themes from the Cairo Geniza have recently been described: some prescriptions attributed to Arnau de Vilanova, a fragment of *materia medica* and some astrological predictions.[55]

To illustrate the parallelism between Moriscos and Marranos and diffusers of technology, I have concentrated on the Marranos in the Islamic world. But it is useful to provide a few indications about the activities of the Spanish Jews in Europe and America as well. As I indicated above with respect to Venice, the Italian states had a tendency to prize the Jews, Spaniards in particular, as (in the words of the Duke of Savoy in 1652) 'inventors and introducers of new arts'. Likewise, Jacobo Moreno asserted that 'after the arrival in our city of Nice of the nation of Israel new arts were introduced and trade enlarged'.[56]

Nevertheless, with respect to Holland, where the densest concentration of Spanish Jews resided, they brought with them (according to Jonathan Israel) 'few techniques besides that of sugar refining'.[57] The association of *conversos* and Marranos in the manufacture of sugar is the main area of Jewish technology in colonial America. A historiographical tradition that began with Antoni de Capmany states that this technique was introduced in Madeira by some Jews expelled from Portugal.[58] From Madeira and Santo Thomé, where Jewish sugar refiners were also active, the Portuguese introduced the technique to Brazil with the participation of Jewish or *converso* technicians or mill owners. Such a technician was Diogo Fernández, brought to Brazil in 1535 by Duarte Coelho as an expert in the manufacture of sugar. From Inquisitorial records, we know the names of quite a few Portuguese Judaizers who owned mills or were identified with the sugar industry in some technical or financial capacity. Many of these persons emigrated to Pernambuco after the Dutch conquest in 1630, where they continued their involvement in the industry.[59] At the end of the century, a nucleus of Dutch Sephardim had established a strong sugar industry in Surinam where, in 1694, forty sugar plantations belonged to Jewish owners.[60]

IV

The problem of the supposed diffusion of fulling mills (*batanes*) from Spain to North Africa and the East by Spanish Muslims and Jews alike again evokes the parallelism of technical diffusion promoted by both groups. In 1622 a group of Tunisian Moriscos was granted a licence to build a diversion dam for a fulling mill on the site of some Roman ruins in the Medjerda River in a place known today as al-Batan. *Baṭan*, pl. *baṭanāt*, is again presumed to be a Spanish introduction, as we noted above in reference to the *shashiya* industry.[61] In the Jewish case, Braude insists that the introduction of mechanical fulling in the Ottoman Empire was owing to Jews from Spain. He describes how, in the environs of Salonica, 300 persons worked in the fulling mills of the Vardar River and its tributaries, and he mentions a half-dozen other centres as well. In Hebrew sources of the sixteenth century the fulling mill is called *batan* and the fuller *batañero*. Braude concludes that 'the Spanish origin of the term ... suggests the western origin of the technology'. Later, the technical terminology entered the Turkish language (*batanciyan*, fuller) and also local Arabic dialects, as in Palestine.[62] There, in Safed, another group of Jewish exiles from Spain established a textile industry at the end of the sixteenth century. The ruins of one of the three documented mills serving the industry survives with the name Tahunat al-Batan (the fulling mill).[63] This was an overshot, vertical mill which had previously been a gristmill.[64]

Braude admits that the etymology of *batán* presents difficulties. Wehr, in his dictionary of modern Arabic, not only identifies the term as a Hispanism, but also limits its usage to Tunisia.[65] The matter is complicated by the fact that *batán* is almost certainly an Arabism in Spanish, from the root *b-ṭ-n*, even if Corominas does not discard a possible Latin derivation from *battuere*.[66]

It is curious and somewhat distinctive of the literature on the exiles of 1492/1615 that its authors focus so closely on terminology. Let us now look at the machine itself. From where did the vertical mill, the fulling mill in particular, come? Here the historical path is suggestive. The horizontal mill appears first in documents from the middle countries of Western Europe and diffuses towards the periphery, while the vertical mill first appears in Southern European documents and diffuses northwards. Furthermore, the horizontal mill was totally dominant everywhere to the east of Syria.[67] For Blaine, these movements suggest that the vertical mill, known to the Romans at least in theory, could have received a powerful stimulus owing to the pre-eminence of the *noria* in the Islamic world, from which it diffused both to the East and North. The fulling mill is a blending of two principles: that of the vertical mill plus the cam, which was also known to the classical world.[68]

In al-Andalus the *noria* reached a high level of development and was widely disseminated in the Christian world, notably to Catalonia. Likewise, the fulling mill, normally presumed of vertical design, could well achieve the same level.[69] Thus we can conclude provisionally: (a) the term *batan* is an Arabism in the peninsular languages; (b) the Arabs, by analogy with the *noria*, were very familiar with the principle of the vertical mill; (c) as inheritors, along with the Europeans, of classical technology they may well have achieved the fusion of the cam with the vertical mill;[70] (d) both the fulling mill and the vertical grain mill diffused from al-Andalus to Castile, where both were called by a synonym of *noria, aceña*;[71] and (e) it was this technique, now expressed in Spanish by the Arabism *batán*, that was equally diffused by Jews and Muslims expelled from Spain.

Joseph Nehama would like a Sicilian origin for the fulling mills of Salonica.[72] Braude admits the universality of the fulling mill in Western European countries and admits the possibility that the universality of Spanish technical terms among Jews in Ottoman lands might have hidden a plurality of routes of diffusion.[73]

V

It is completely logical that folk migrations produce as a natural corollary the transmission of ideas and techniques. In the history of technology, diffusion, whether by migration or by some other means, has been a typical stimulus to innovation. The process is as characteristic of traditional societies as it is of industrialized ones.[74] The interpretive problem comes from the fact that, even though in some cases the Ottomans were technically inferior to the Europeans, they were at the same time possessors of a culture with a high tradition in the practice of science and technology. The Turks wanted to take advantage of those things they lacked and, therefore, the Hispanicity of Marranos and Moriscos was attractive to them. In this sense the familiarity of both groups with the new American flora is an obvious symbol.

On the other hand, in the case of Marranos in Holland or Italy, one cannot allude to any technical inequality between those countries and sixteenth-century Spain. In this case, therefore, the logic supporting any significant technological displacement does not exist. Here, the myth of the technical skill did not rest on the Hispanicity of the Marranos but rather on the supposed scientific superiority of the Jews.

Nevertheless, the fact that the same technologies were attributed to both cultural groups, and by very different and mutally isolated historical traditions, is a proof of the reality of the process.

Notes and References

1. Giovanna Vitelli, 'Andalusians in Medieval Tunisia: the Determinants of Culture Contact and Change', in T. F. C. Blago, R. F. J. Jones and S. J. Keay, eds, *Papers in Iberian Archaeology* (Oxford, 1984), 704–26, on 705.

2. Américo Castro, *The Spaniards* (Berkeley, 1971), 298. See my comment on this passage in *Islamic and Christian Spain in the Early Middle Ages* (Princeton, NJ, 1979), 246–7.

3. Pierre Ponsot, 'Les Morisques, la culture irriguée du blé, et le problème de la decadence de l'agriculture espagnole au XVIIe siècle', *Mélanges de la Casa de Velázquez*, 1971, 7: 239–62.

4. Muhammad Razouk, *Los moriscos y sus migraciones a Marruecos durante los siglos XVI y XVII* [Arabic] (Casablanca, 1991), 266. Al-Shatibī means 'from Játiva'.

5. *Ibid.*, 188. *Taba* was a colloquial expression, instead of the more correct form, *tabaq*.

6. Louis Cardaillac, 'Morisques en Provence', in M. de Epalza and R. Petit, eds, *Études sur les moriscos andalous en Tunisie* (Madrid, 1973), 89–102, on 97–8. On these drugs, see José Pardo Tomás and María Luz López Terrada, *Las primeras noticias sobre plantas americanas en las relaciones de viajes y crónicas de Indias (1493– 1553)* (Valencia, 1993), 210 (*mechoacán*) and 206–7, 308 (*ocozotl*).

7. J. D. Latham, 'Towards a Study of Andalusian Immigration and Its Place in Tunisian History', *Les Cahiers de Tunisie* 1957, 5: 203–52, notes 276–80. See also E. Gobert, 'Les références historiques de nourritoures tunisiennes', *Cahiers de Tunisie*, 1955, 3: 501–42, on pp. 529 (chili pepper), 531 (Opuntia).

8. Juan Vernet, *El Islam y Europa* (Barcelona, 1982), 146.

9. Gobert, *op. cit.* (7), 529.

10. Latham, *op. cit.* (7).

11. Ahmed Kassab, 'L'evolution d'un village "andalous": Testour', in *Etudes, op. cit.* (6), 359–68, on 367.

12. Muhammed El Aouani, 'A la recherche des influences andalouses dans les campagnes tunisiennes: Essai de mise au point', in *ibid.*, 374–7, on 377.

13. Hafedh Sethom, 'L'apport andalous à la civilisation rurale de la presque'ile du Cap Bon', in *ibid.*, 369–73, on 372.

14. See Ponsot, *op. cit.* (3).

15. Razouk, *op. cit.* (4), 288–90.

16. Richard W. Bulliet, *The Camel and the Wheel* (Cambridge, MA, 1975), 20, 26, 90–1.

17. *Ibid.*, 202. The wheeled cart also must have survived in al-Andalus, where there are archaeological remains of wheel ruts carved into roads during Islamic times. See Manuel Riu, 'Aportación de la arqueología al estudio de los mozárabes de al-Andalus,' in Riu *et al.*, *Tres estudios de historia medieval andaluza*, 2nd edn (Córdoba, 1982), 85–112, on 92–3, where he says that access roads to Mozarab village sites preserve the imprints of cart wheels, permitting calculation of the width of axles and even wheel diameters.

18. Bulliet, *op. cit.* (16), 204.

19. *Ibid.*, 204–5.

20. Latham, *op. cit.* (7). In note 208, he criticizes M. Rodinson, who presumes an Italian origin for *kirrīta*: 'Adjala', *Encyclopaedia of Islam*, 2nd end, I, 206. But Rodinson also assumes, like Latham, that the term only entered North Africa in the sixteenth century.

21. Razouk, *op. cit.* (4), 289.

22. Mercedes García-Arenal, 'Los andalusíes en el ejército Sa'dí: un intento de golpe de estado contra Ahmad al-Mansūr al-Shahabī (1578)', *Al-Qantara*, 1984, 5: 169–202, especially 183 and 191, note 66.

23. Bernard Lewis, *The Muslim Discovery of Europe* (New York, 1982), 224.

24. David James, 'The *Manual de Artillería* of al-Ra'is Ibrāhīm b. Aḥmad al-Andalusī with Particular Reference to its Illustrations and their Sources', *Bulletin of the School of Oriental and African Studies*, 1978, 41: 237–57. Ibn Ghānim was born in Nawlash, a fortress in the Valley of Lecrín, today called Niguelas. See James, 250, note 38; Luis Seco de Lucena, *Topónimos árabes* (Granada, 1974), 63–4. His family avoid the Morisco expulsion of 1571 (to which he refers as *khurūj al-awwal*, the 'first departure'); rather they left for Seville in 1584.

25. See the passage on the use of the rule to adjust the sight of a gun, James, *op. cit.* (24), 240–1; also Vernet, *op. cit.* (8), 146.

26. Vernet, *ibid.*

27. Latham, *op. cit.* (7).
28. Muhammad Al-Annabi, 'La chechia tunisienne', in *Etudes* (6), 304–7; Latham, *op. cit.* (7); and, especially, Paul Teyssier, 'Le vocabulaire d'origine espagnole dans l'industrie tunisienne de la chechia', in *Etudes*, 308–16.
29. Teyssier, *op. cit.* (28), 310–11.
30. Razouk, *op. cit.* (4), 289.
31. George Saliba, 'The Function of Mechanical Devices in Medieval Islamic Society', *Annals of the New York Academy of Sciences*, 1985, 441: 141–51, on 142.
32. See Saliba's translation, *ibid.*, 145.
33. Joan Corominas, *Diccionario crítico etimológico castellano e hispánico*, six volumes (Madrid, 1980–91), s.v., 'máquina'. By a process of semantic differentiation, *ingenio* continued to be used with the restricted sense of a sugar mill.
34. See *Oxford English Dictionary*, s.v. 'machine'.
35. Benjamin Braude, 'The Rise and Fall of Salonica Woolens, 1500–1650: Technical Transfer and Western Competition', *Mediterranean Historical Review* 1991, 6: 216–36, on 223.
36. Castilian: *pelaire, perayre.*
37. Normative Castilian: *betaldar.* See Luis Márquez Villegas, *Un léxico de la artesanía granadina* (Granada, 1961), 84.
38. Castilian: *tendedor.*
39. Cited by Braude, *op. cit.* (35), 225. The Hispanisms in the Hebrew text are italicized.
40. *Ibid.*, 225: Eliezer Bashan, 'The Rise and Decline of the Sephardi Communities in the Levant – the Economic Aspects', in R. D. Barnett and W. M. Schwab, eds, *The Sephardi Heritage*, II (Grendon, England, 1989), 349–88, on 361.
41. Cited by Fernand Braudel, *The Mediterranean and the Mediterranean World in the Age of Philip II*, two volumes (New York, 1972–3), II, 808.
42. *Viaje de Turquia*, Fernando G. Salinero, ed. (Madrid, 1980), 428.
43. Fawzi Abdulrazak, 'The Kingdom of the Book: The History of Printing as an Agency of Change in Morocco Between 1865 and 1912', unpub. doctoral diss., Boston University, 1990, 76, 99; Salo W. Baron, *A Social and Religious History of the Jews*, XVIII (New York, 1983), 225.
44. Jonathan I. Israel, *European Jewry in the Age of Mercantilism, 1500–1750* (Oxford, 1985), 181.
45. Salvatore Foa, *La politica economica della casa Savoia verso gli ebrei dal sec. XVI fino alla Rivoluzione Francese* (Rome, 1962), 43.
46. Anita Engle Berkhoff, 'A Jewish Glass-blower from Spain', *Miscelanea de Estudios Arabes y Hebraicos*, 1965–6, 14–15: 43–61.
47. Julio Samsó, 'A propósito de dos libros recientes sobre las relaciones culturales entre España y Túnez', *Ethnica*, 1975, 9: 243–54, on pp. 252–3.
48. Vernet, *op. cit.* (8), 145.
49. Bernard R. Goldstein, 'The Hebrew Astronomical Tradition: New Sources', *Isis*, 1981, 72: 237–51.
50. Ala Marcova, 'El Tratado de astrolabio de Moshe Almosnino en un manuscrito de Leningrado', *Sefarad*, 1991, 51: 437–46: 'estorlabio . . . en romance castellano; desproveído de libros'.
51. Voltaire, *Histoire de Charles XII*, book V, cited by Harry Friedenwald, *The Jews and Medicine*, two volumes (Baltimore, 1944), II, 725.
52. Eleazar Gutwirth, 'On the Hispanicity of Sephardi Jewry', *Revue des études juives*, 1986, 145: 347–57, on pp. 349–52. On Lobera de Avila, see Francesc Bujosa, 'Luis Lobera de Avila', in J. M. López Piñero *et al.*, eds, *Diccionario histórico de la ciencia moderna en España*, two volumes (Barcelona, 1983), II, 529–31.
53. Uriel Heyd, 'Moses Hamón, Chief Jewish Physician to Sultan Suleyman the Magnificent', *Oriens*, 1963, 16: 152–70, on 168–9; Viaje de Turquia, 208. On the medical dynasty of the Hamóns in Turkey, see Freidenwald, *op. cit.* (51), 729. Isaac Hamón was physician to the Nasrid sultan of Granada in 1475; Rachel Arié, *L'Espagne musulmane au temps des Nasrids*, 1232–1492 (Paris, 1973), 334.
54. *Viaje a Turquia*, 170.

55. E. Gutwirth, 'Judeo-Spanish Fragments from Cairo', *Anuario de Filología*, 1983, 9: 219–23, on 220. None of these three fragments can be dated with certainty, but all circulated in the Ottoman world.

56. Foa, *op. cit.* (45), 43 (letter from Carlo Emmanuele II, 2 December 1653): 'Gli inventori et introduttori di nove arti et traffichi ... dopo la venuta della natione hebrea nella città nostra di Nizza si sono introdotte nove arti et accresciuto il commercio.'

57. Jonathan I. Israel, 'The Economic Contribution of Dutch Sephardi Jewry to Holland's Golden age, 1595–1713', *Tijdschrift voor Geschiedenis*, 1983, 96: 505–35.

58. Arnold Wiznitzer, 'The Jews in the Sugar Industry of Colonial Brazil', *Jewish Social Studies*, 1956, 18: 189–98, on 190. The date given by Capmany (1549) is clearly too late.

59. *Ibid., passim.* Also Wiznitzer, *Jews in Colonial Brazil* (New York, 1960), 67–70.

60. Israel, *op. cit.* (44), 177.

61. Latham, *op. cit.* (7). Also Teyssier, *op. cit.* (28), 310.

62. Braude, *op. cit.* (35) 224–5.

63. *Tahuna* = mill. The Arabism *atahona, atafona*, in the sense of a hydraulic mill, entered Castilian in the thirteenth century; Eero K. Neuvonen, *Los arabismos del español en el siglo XIII* (Helsinki, 1941), 249–50.

64. Shmuel Avitzur, 'The Batan: A Water-powered Fulling Mill in Nahal Ammud', *Israel – Land and Nature*, 1981, 7: 18–21.

65. Hans Wehr, *A Dictionary of Modern Written Arabic* (Ithaca, NY, 1976), 64.

66. Corominas, *op. cit.*, I (33), 541–53.

67. Joseph Needham, *Science and Civilisation in China*, IV, part 2, 'Mechanical Engineering' (Cambridge, 1965), 368.

68. Bradford B. Blaine, 'The Enigmatic Water-Mill', in B. S. Hall and D. C. West, eds, *On Pre-Modern Technology and Science: Studies in Honor of Lynn White, Jr* (Los Angeles, 1976), 163–76. On the *noria* as a paradigm for all geared mechanics, see Glick, *op. cit.* (2), 238.

69. I omit a consideration of the possible Chinese origin of the *batan*. The Chinese had devised a method of using a horizontal mill to work mallets, with pulleys, even though this task is more efficiently handled by a vertical mill (Needham, *op. cit.* (67), IV.2, 373–4, 390). I mention this because in medieval Valencia there were horizontal fulling mills, although we have no idea of how they worked.

70. According to Lynn White Jr, that fusion probably took place in some centre to the north and east of the Roman Empire which remains to be identified: *Medieval Technology and Social Change* (Oxford, 1962), 81; Blaine, *op. cit.* (68), 170, puts the union of the vertical mill and cam in the Alpine region in the ninth or tenth century.

71. On the *aceña* as a vertical mill, see Glick, *op. cit.* (2), 231.

72. 'The Jews of Salonica in the Ottoman Period', in *The Sephardi Heritage*, II, 208.

73. Braude, *op. cit.* (35), 228–9.

74. On the migration of artisans, see Glick, *op. cit.* (2), 221–2.

Variations on Mass Production: The Case of Furniture Manufacture in the United States to 1940

CARROLL PURSELL

Historians of American industry agree that classical mass production, such as that perfected in Detroit's automobile industry, never characterized the manufacture of furniture in the United States.[1] As one historian has put it, 'there was no Henry Ford of the furniture industry'.[2] At the same time, however, as Philip Scranton has recently warned us, we should not let the great success of, and subsequent focus on, mass production industries blind us to the interesting and significant range of what he calls 'batch' rather than 'bulk' industries, among which he lists furniture.[3]

Even a cursory investigation of furniture manufacture reveals that it reached a stage, in a few atypical factories, very like mature mass production years before the birth of the automobile industry. After the success of Henry Ford, the idea of mass production existed as an ideal for many furniture manufacturers, and indeed significant aspects of the mass-production system were introduced into furniture factories. And finally, with the introduction of 'modern' furniture, new materials and fabrication techniques came into the industry, changing parts of it dramatically. The result of these introductions was the strengthening of what Scranton calls 'flexible' rather than 'mass' production.

Until the end of the eighteenth century, furniture making remained almost entirely a handcraft occupation. The great names of Chippendale and Hepplewhite had used hand tools with an artistry unsurpassed, and produced fine pieces of furniture at prices few could afford. At the same time, country artisans crafted pieces from local woods, with hand tools, for local (often personal) use. Then in a burst of inventive genius almost unparalleled in history, Samuel Bentham in England patented, in 1793, what one twentieth-century mechanical engineer called 'practically every woodworking machine and process that is in use today'.[4] Bentham designed his machines for use in the Royal ship-building yards and in prisons, in which his brother Jeremy Bentham was much interested. Broad scale use of his machines came eventually, though only slowly, to the woodworking industry.

In the United States, by contrast, where an abundance of wood made its working a major industry, additional machines were introduced, such as the famous Blanchard copying lathe used to make gunstocks in the national armouries.[5] By 1853, official British observers of American industry expressed their admiration for the quality and number of American woodworking machines, and the fact that they were so widely used. 'In no branch of manufacture does the application of labour-saving machinery produce by simple means more important results than in the working of wood', wrote Joseph Whitworth.[6] By that time William T. Powers, for example, a cabinet-maker 'with an especial liking to machine work pertaining to the cabinet business', had built a large furniture factory in Grand Rapids, Michigan, a town whose name was to become synonymous with the industry. In 1851 his partner was able to write that 'we can almost as it were throw in whole trees into the hopper and grind out chairs ready for use'.[7]

At mid-century, Cincinnati, Ohio, was the centre of the western furniture industry, and some plants located there were turning out a wide variety of products in very large numbers. One firm, specializing in bedsteads, offered over 95 varieties. Another firm, perhaps also a specialist, was turning out 180,000 chairs annually. Yet another company, which worked out of a seven-storey building, provided 30,000 chairs, of different types, annually to just one buyer in St Louis.[8] Cincinnati, of course, was also the centre of wagon and carriage making and of the machine tool industry in the west.

These statistics from Cincinnati suggest two important facts about the furniture industry in relation to mass production. First, the fact that one firm would provide 95 varieties of bedstead stands in dramatic contrast to Henry Ford's later conceit that customers could have any colour Model T they wanted as long as it was black. The flexibility to produce so many different designs was hardly compatible with mass production as Ford envisioned it. Second, the very large number of chairs produced suggests that they were not simply being hand crafted by artisans. It suggests, rather, that some system, incorporating woodworking machines but not necessarily limited to them, was in place. A study of hand versus machine labour in the industry, published in 1899 by the United States Commissioner of Labor, claimed that 'machinery has revolutionized the woodworking industry', but also complained that when trying to compare methods in the furniture branch, 'owing to the large number of different articles represented under this classification, [and] the various methods and machines used for similar work', comparison was very difficult indeed. 'When it is remembered that the prevailing styles of furniture made at the dates of production of most of the hand units have been substituted by others', the report explained, 'it will be seen how difficult it has been to procure data for units exactly alike under the two methods.'[9] Still, as one historian of the Grand Rapids industry has written, 'by 1870, the machine was clearly triumphant, and mass production of furniture for mass markets was fully possible'.[10]

David Hounshell has given us an example of furniture production that came very close to that ideal. For the first decade of its existence the Singer Manufacturing Company, the leading producer of sewing-machines in the United States, bought wooden cases for its machines on contract from New York cabinet-makers. Then in 1868 it moved that part of the business to a new factory it built in South Bend, Indiana, which was joined in 1881 by another plant constructed in Cairo, Illinois. So enormously successful were these plants that by 1920 that in South Bend alone was turning out more than two million sets of wooden cabinets a year.[11]

During the 1880s the technology of Singer's operations was revolutionized by the adoption of cheaper wood (gum) for its cabinets, its use of plywood (built-up layers of veneer) components and its adoption of formed, or bentwood, pieces. Hounshell has discovered that 'abundant evidence documents the adoption at Cairo and South Bend of increasingly specialized machines and tools for working wood, many of which were built in Singer's [own] ... works'.[12] Hounshell further suggests, however, that while improved machinery may have been a necessary cause of Singer's very large production, it was almost certainly not a sufficient cause. Rather, 'the nature of the product and the market' were also of critical importance. Cabinets were not sold on their own, but as an almost necessary accompaniment to sewing-machines. To the extent, then, that Singer dominated the sewing-machine market, it had a ready and almost guaranteed market for its cabinets. It was an advantage that chair and bedstead makers did not have.[13] In the industry as a whole, styles changed, and with them consumer taste. Production runs sufficiently large to justify true mass production were difficult, if not impossible, to make.

What this indicates is not that the furniture industry somehow failed, or remained a backwater, mired in artisanal methods. Instead it indicates that the industry falls in that group called by Philip Scranton 'custom/batch' rather than 'bulk/mass'. As he points out, 'the choice between variety and volume had strong technical correlates, built into the structures of production': most specifically, industries of the custom/batch variety made heavier use of general, multi-purpose tools and relied less on dedicated, single-purpose tools.[14] Tools also tended to be arranged differently in the two types of industry, by function in the batch and in order of production steps in mass. And to cap it off, batch manufacturers tended to be more dependent on workers who, like the machines, were general purpose: skilled craftspeople who had the flexible knowledge to shift jobs as needed.

As Scranton recognizes, production is only one part of the industrial system, whether in batch or mass-production firms. In the former, skilled and knowledgeable salespeople were especially important as mediators between factory and customer. Marketing relied more upon catalogues, branch offices, dealers and travelling salespeople. 'Product diversity, market fluctuations, dependence on skilled labour', Scranton writes, 'all characterized the practices of participants in the batch production division of American industry, despite their varied scales and settings.'

It would be a mistake, of course, to draw the boundary between these two types of industry too rigidly. Scranton notes that 'batch and custom efforts intersected with bulk manufacturing at virtually every step in the creation of the nation's transportation, power and communications system'. At the same time, however, batch industries borrowed both the ideal and many of the techniques of mass production, installing those which seemed to make sense within the constraints of flexible, fashion-driven production. The furniture industry clearly fits the pattern of Scranton's custom/batch production, but also demonstrates the permeability of the boundary between the two domains of production.

Despite the high degree of mechanization in the industry, members of the American Society of Mechanical Engineers (ASME), inspired by the triumph of Henry Ford's method of mass production and Frederick Winslow Taylor's system of 'scientific management', made a concerted effort to 'modernize' American furniture making. In 1920 the manager of a Michigan veneer works asserted in disgust that 'woodworking, one of the oldest civilized trades, is now one of the largest industries in the United States. It is doubtful whether any group of modern manufacturers gives evidence of less scientific knowledge of its products.'[15] The next year ASME member Byron Parks, also of Michigan, took part in a special session at the society's annual meeting organized by the Committee on Woodworking.

In a paper titled 'Engineering in Furniture Factories', he declared that 'the woodworking industry . . . has shown the least development and has been the slowest to adopt modern principles of manufacturing of any industry of which the writer has knowledge'. His condemnation of the industry was virtually complete. He denounced 'the general lack of accurate cost data and the absence of technically trained men in executive positions'. Indeed, he claimed not 'to recall a single man with a technical education or training'. He was convinced, he wrote, 'that the average manager of a furniture plant is more interested in marketing his product than in manufacturing it'.[16]

Parks decried the lack of technical people in the factories, but significantly listed 'skilled labour' as one of the assets of any area where furniture factories were concentrated. Grand Rapids, Michigan, he wrote, had originally been close to its supply of lumber. Then, 'as the number of plants multiplied a good supply of skilled labor was accumulated, so that now, even when Grand Rapids is not so well situated as other centers as regards nearness to raw-material supply, the labor market as well as a certain reputation attaching to Grand Rapids products has tended to increase the number of furniture-manufacturing plants' even further.[17]

The complaints and dreams of rationalization of such engineers as Parks led, in 1925, to the establishment of a formal Wood Industries Division within the American Society of Mechanical Engineers.[18] Progress towards their ideal of mass production, however, was slow. In 1929 the President of the Tomlinson Chair Manufacturing Company in North Carolina proudly characterized his own firm in terms that confirmed the worst fears of Parks and his associates, but sound more familiar to those accustomed to the recent claims of 'lean production'. 'So important has the style element in furniture merchandising become', he claimed, 'that the successful manufacturer today is the one who has abandoned old manufacturing theories.'[19] 'We have found', he went on, 'that we must view our entire manufacturing business from a sales and merchandising point of view.'

In-house stylists and designers kept 'their fingers on the pulse of the great Eastern market'. In addition they had 'a field man travelling throughout our sales territories, who is constantly surveying retail markets'. When a new design is decided upon,

> a factory detail is prepared and released to the planning department which has nine men [as compared with 700 in production] and is the nucleus of our manufacturing activities. Specifications and routings are made out and turned over to the sample department for the making of models. A completed model is made for each new design and this must be approved by a merchandising committee before a production order is issued. The sample department is a completely equipped shop in itself, specializing entirely on sample work.

The variety of the firm's products was staggering. 'We have possibly 1,200 different designs', Tomlinson claimed, 'most of which are changing rapidly. Our turnover of designs is large.' It is not surprising, therefore, that the mass-production ideal of the engineers was not even attempted. 'Many of our production runs', according to Tomlinson,

> are comparatively short. This requires the ability to make set-ups quickly, and demands good dispatching. It would be more economical from a manufacturing standpoint to run larger lots in many cases, but we are governed by sales requirements, and we do not make more units than we believe we can sell.

Significantly, Tomlinson ended by claiming that

> our manufacturing organization is very enthusiastic about the new order of things. Work is more interesting. Initiative is in demand. An opportunity is given to see what can be done. At our weekly foreman's meeting on Monday nights we plan constructively.

Small production runs, however, did not always mean that production techniques could not be improved. The handling of materials and products was a critical part of any manufacturing effort, and some factories were carefully planned from this standpoint. H. K. Ferguson, president of a company in Cleveland, Ohio, that made handling equipment, claimed that a large plant he knew of making office equipment had created a miracle of efficiency through better planning and equipment. 'The installation of mechanical-handling equipment in one department', he boasted, 'saved 10,000 square feet of floor space, cut the inventory in that department 80% and added a third to the output, with a $59,000 cut in labor costs.'[20]

Another woodworking plant cited by Ferguson, 'where spokes and wheels are made, was completely rearranged, and mechanical handling equipment was installed. Twenty-six men were taken out of each line of production with very little increase in investment for equipment and upkeep.' The type of equipment his own company sold, overhead cranes for example, were power-driven aids to handling, but Ferguson also advocated the creative use of gravity. One mid-western furniture plant he mentioned had a factory floor laid on a 0.5 to 1 per cent grade, so that one end of the building was 4 feet higher than the other. 'Practically a straight line of flow was developed from the green lumber to the shipping tracks for a distance of over 1,000 feet. It is this sort of layout which simplifies the material-handling problem and keeps down the cost of operation.'

The First Vice-President and Production Manager of the Showers Brothers Company, a furniture firm in Bloomington, Indiana, also addressed the question of material handling, but placed it in a larger context of production. Significantly, the firm manufactured furniture towards the low end of the price range. 'Were we manufacturers of furniture of the most expensive type, I suppose we might go ahead and build it in the best possible way and let the price fall where it will.' At his end of the product line, however, trying to 'keep the price down within the reach of the modest pocketbooks' required, as he said, 'cutting manufacturing corners ... establishing quicker and better ways of doing the same old thing'.[21]

'It all boils down', he wrote, 'to quantity production, to straight-line manufacturing methods ... It has become almost a fetish with every member, from works manager down.' The captions of the photographs which illustrate his article tell much of the story:

> Belt-drive, ball-bearing roller conveyors speed the flow of materials from cut-off saw to rip-saw, and on to the next operation.
>
> No unnecessary rehandling in the dimension mill. Compact arrangement of machines and conveyors saves several men's energy.
>
> From rough lumber to drawer-ends in four steps. That means cutting manufacturing corners. Here is a machine that moulds, planes, and sands four sides, then grooves the board, all in one operation.
>
> Turntables, located in the kitchen cabinet assembly line, enable the workmen to twist the case into a convenient position. Time-saving power tools are used wherever possible to keep manufacturing costs at a low figure.
>
> Assembling on a runway of convenient working height. Each man has a job to do. One puts on hinges, another molding, a third fits drawers, and so on down the line of workers.

Some of the new machines came from the shops of the furniture plant itself. While many of them were 'standard to wood-working practice', others 'have been brought out by our own men'. The

> units of furniture go through in lots of 200 to 400, and to insure a continuous production flow, machines of the same kind or of a similar nature, such as band-saws, planers, or shapers, have been grouped in blocks, which are under the close personal supervision of the foreman. The men in each block develop a real pride in the work and put forth every effort to keep jobs moving at the highest rate of speed consistent with quality workmanship.

Significantly, the last photograph caption reads, 'Spindle carving calls for steady nerves and a skill that years of service bring.'

The author did not say where this 'fetish' for 'straight-line manufacturing methods' came from, but he did provide one hint.

> Ready now for assembling, which requires a lot of little operations, we have taken a leaf out of the automotive industry's book and do this work on a conveyor, each man having one particular job to do.

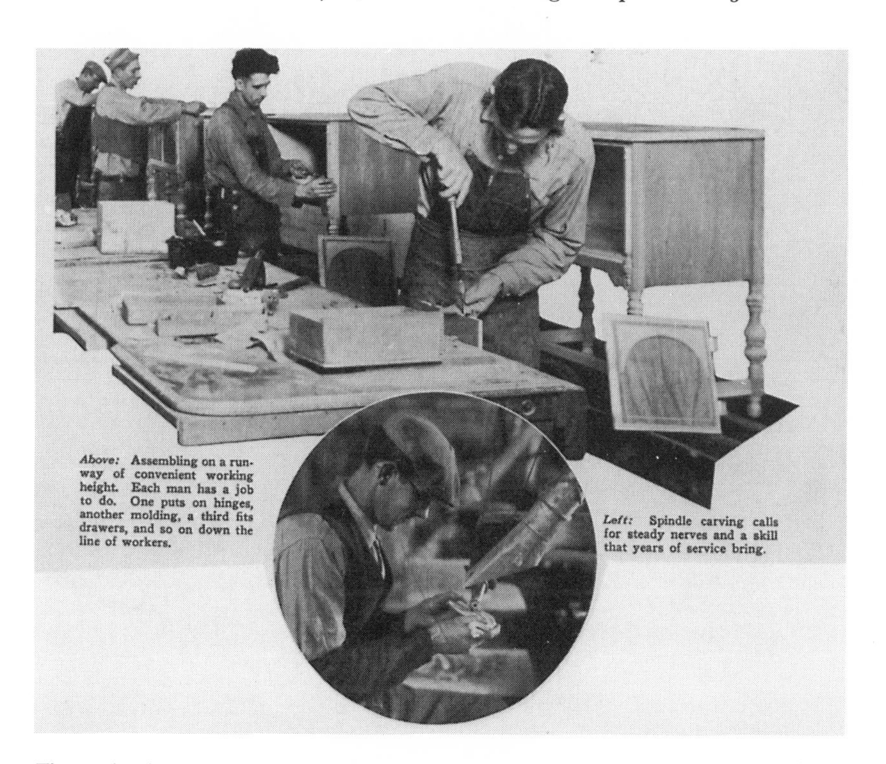

Figure 1 Assembly line runway.
Source: From C. A. Sears, 'Lower Costs by Down-hill Handling', *Factory and Industrial Management*, 1928, 76: 695.

> Our conveyor differs from the kind found in automobile factories, however. It isn't power-driven.

The reason wasn't mysterious. 'I question whether a power-drive conveyor would be practicable', he wrote. 'It would certainly be risky business to fit a door on a cabinet moving however slowly.' Conveyor, Sears admitted, was perhaps not the right word for his installation: in fact it was 'a long, waxed runway of convenient height'.

In 1930 M. C. Phenix, General Manager of the L. C. Phenix Company in Los Angeles, described the operations of his own plant, which seem to have taken the automobile analogy about as far as possible in a furniture factory. Again, the flavour of the effort is given in a photograph caption: 'All it takes to make an upholsterer in a line-assembly system is that he know how to drive tacks. From that point we can readily teach him in a few days.'[22]

'For years', Phenix wrote, 'both my father and I had dreamed of putting our furniture factory on a line-production basis.' One problem was that they 'were manufacturing 150 to 200 different individual styles of upholstered furniture in our plant'. They were, as he admitted, 'custom manufacturers in practically every sense of the word'. The dream, however, came true, at least in his own estimation. 'Since then', he claimed,

> we have proved conclusively that line production is as practicable for the manufacture of upholstered furniture as it has been in many other industries to which it has been applied. We have effected savings that in some instances have been phenomenal. We have

Figure 2 Upholstery assembly line.
Source: From M. C. Phenix, 'Line Production', *Factory and Industrial Management*, 1930, 80: 1170.

> speeded up production and injected into our plants efficiencies that were unheard of under the old plan of assembling.

The first move, not surprisingly, was to cut the number of styles offered, from between 150 and 200 to five 'primary styles', though, somewhat mysteriously, he admitted that this had made it necessary, 'for competitive reasons', to 'make a duplicate line'.

'A good deal in the way of shop practice', Phenix asserted, 'was borrowed from other industries which have used line production with good results.' As always, the new practice was only in part mechanical. The firm adopted the piece-work system of paying workers, and, 'we abandoned our former plan of working against orders on hand, and started working against a production schedule'. Further, stocks and supplies were ordered automatically when supplies on hand dropped below a specified level and costs were figured against production at the end of each day. 'Electric drills and other electric tools are used to speed up production without lowering the quality of the product', new spraying techniques allowed the application of finishes at one-tenth the previous cost and all chairs, for example, moved from the fourth to the first floor through gravity chutes.

The signature of mass production was, of course, the assembly line, and here too Phenix borrowed from the automobile industry. 'There are seven men on the davenport line', Phenix wrote. 'These men turn out one and one-half times as much as we produced under the old system with the same number. They stay in position at all times, the stock being brought to them by stock boys serving each line.' The technical details of the 'line' are not revealed, but Phenix quite specifically reveals that its impact on the labour force is parallel to that in the automobile industry. 'Custom methods of furniture manufacturing', he brags, 'require a great deal of upholstering knowledge on the part of each man. All it takes to make an upholsterer for a line-assembly system is that he know how to drive tacks. From that point', he concludes, 'we can readily teach him in a few days.' In conclusion, Phenix claimed that 'it now takes a two-piece suite three hours to go through from the parts stock to the finished product. Under the old plan, it took from two to three days.'[23]

The piecemeal introduction of elements of mass production from the automobile industry represented only a part of the modernization of furniture manufacture in twentieth-century America. The changes described all took place within the production of traditional furniture, made of wood with perhaps some fabric upholstery. By and large these products were, themselves, little changed.[24] At the same time, however, new designs made their appearance with the advent of 'modern' furniture, new materials such as steel, aluminium and plastics were more widely used, and segments of the institutional market, such as schools, businesses, government offices and hotels, greatly expanded.

The increasing use of steel and aluminium in furniture was closely tied to the growth of the institutional market. One economic historian wrote

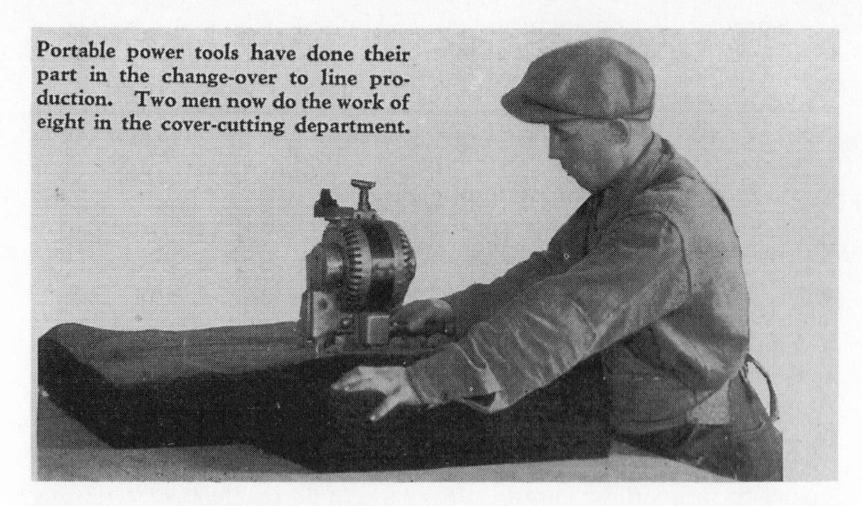

Figure 3 A portable power tool.
Source: From M. C. Phenix, 'Line Production', *Factory and Industrial Management*, 1930, 80: 1171.

at the end of the 1920s that 'probably more household, schools and office furniture was manufactured in the United States during this period than in any other country in the world'.[25] The increasing size of both business and government in the country, coupled with their increasing bureaucratization, led to a strong and growing demand for office furniture. The concomitant increase in the size, number and opulence of hotels and department stores, and the growing number of children attending schools, also created a large demand not only for chairs, couches and so forth, but also for desks and bookshelves. It was mainly to meet these institutional markets that metal furniture was produced.[26]

It was reported in 1917 that although until that time most metal furniture had been built to order, an increasing number of firms were beginning to do 'stock work', and were even beginning to specialize. 'Standardization of construction has lately been adopted by many concerns', one observer noted, 'but it is a slow process on account of the alteration or betterment of construction that is continuously being made in order to reduce the cost of production.' Whether cause or effect, it was also noted that 'few engineers and draftsmen have shown interest in metal-furniture manufacture'.[27]

The lack of standardization, owing in part to changing methods, led this observer to declare that 'so many manufacturing methods are employed that it is impossible to state which is preferable', a lament which had been heard a century before when European visitors remarked on the same condition of American railways. In the case of furniture, it was claimed that 'the field is open to the inventor'.[28]

One obvious technique, perhaps borrowed from the automobile industry, was electric welding, which was described as having 'revolutionized the metal-furniture industry by reducing the amount of labor necessary'. Moreover, 'in the press room the plates are divided into classes, so that it will not be necessary to change the dies or machines too often'. Most interesting was the handling of the problem of a gap between the factory drawings of parts, and the types of materials used. Since both steel sheets and tubing were often bent into shape, one needed to know how to adapt the drawings to the dimensions of the steel used. 'Until the adoption of definite allowances for bends, which were published in *Machinery* in 1911 and 1915', a writer in that journal declared,

> each foreman was the ruling power in his department, and all the young engineer or draftsman could say was, 'Oh, yes; I see,' although he never did. As it would greatly increase the expense if the manufacturer had to have a complete mechanical drawing made in each case, it has been left to the listers of materials, who generally are practical men, to check up the construction as to practicability, etc.[29]

The working out of such practical handbook formulae was a practical necessity and an important step in standardizing the manufacturing processes.

A production engineer describing the manufacture of 1,200 steel chairs for government hospitals in 1918 revealed that oxyacetylene as well as electric welding was used. 'Cast-iron forming dies' were used in bending the tubing, then jigs were made to hold the parts together for the necessary 12 acetylene welds. As in many other industries entering mass production, a great deal of handwork remained. 'Since even the best acetylene welder can not make a weld smooth enough to enamel over without being finished', he admitted, 'there was considerable filing to be done.'[30]

In 1923 the Simmons Company was embarked on the manufacture of bedroom furniture from steel. They offered nine distinct suites, each with a possible eleven pieces, producing 300 separate pieces a day, a rate which was expected to rise tenfold as soon as a new plant, then under construction, was ready to go. Again the pieces were fabricated from seamless tubing, electrically welded. Since 'an important feature of steel construction which should not be overlooked is that the cheapest bedroom suite differs not one iota from the most expensive in strength and durability', it was observed, the problem of product differentiation was extreme. It was solved by painting the suites at the upper end with an imitation wood grain. This work was entrusted to 'workmen of great skill', since it was done 'entirely free hand with the consequence that the work entirely lacks the artificial appearance of a printed or stenciled grain'.[31] Like hand filers, hand painters marked a significant, and expensive, gap in modern production methods.

Tracing the development of the industry, a writer for *The Iron Age* in 1935 declared that metal furniture was 'unheard of 40 years ago'.[32] Then,

> almost concurrent with the development of the automobile, inquiring minds closely associated with the steel industry, ever on the scent for new product outlets, were the first to bring out a line of filing cabinets, desks and other furniture. The initial effort was confined to office furniture; the home was not invaded until later.

This early furniture, it noted, was made to look like wood, 'stressing the limitations of metal rather than its advantages'.

Like steel, early aluminium furniture was used first for offices. In 1923, according to one story, Richard B. Mellon, who from 1899 to 1910 had been President of the Aluminum Company of America, decided to have a swivel chair made of this material for his office at the Mellon National Bank in Pittsburgh. This first chair had been cast, but as orders began to roll in for similar pieces – 100 desks and 300 chairs for the new Mellon Bank building, 700 chairs for the Free Library in Philadelphia, 1,200 chairs for the dining cars of the Pennsylvania railroad – they were soon made with sheet aluminium and tubing. It was still true, however, that these early pieces were virtually handmade.[33]

The Aluminum Company had always considered itself a producer of metal for fabrication into consumer goods by other firms, but in order to follow up on the interest in furniture it decided to enter that business directly. We are informed that

> part of the Buffalo works, where aluminum automobile bodies were being made, was turned over to the furniture division. Production and sales organizations were assembled, jigs and tools were manufactured for a representative number of designs and a small stock was accumulated. As soon as regular production could be established, the prices of pieces dropped sharply.

Following the steel example, office furniture at first dominated, but then the 'line was widened to include items adaptable to hotels and restaurants, schools and colleges, hospitals and kindred institutions'.[34] It is clear that the identification of aluminium with modernity, and its relative ease of working, led to the wedding of modern design with furniture fabricated from this material. The fact that 'almost every hotel that was built ... had to have at least one public room in which aluminum furniture was the outstanding piece of decoration' was surely a sign of its modernity. Before the Aluminum Company turned over its fabrication business to the General Fire-proofing Company of Youngstown, Ohio, 'leaders of modern art, to which aluminum was peculiarly adapted, took up the design of aluminum furniture'.[35]

The logic of designing furniture in modern styles and of metal made some sense in terms of the constraints of mass production, but not until the mid-1930s could *Business Week* claim that 'furniture makers and sellers, see [a] real shift to modern design'. In an article titled 'We Go Modern', three problems were identified as having held back the movement until then. First, an early flood of cheap, poorly made 'modernistic' furniture had soured the market for the real thing. Second, the American buying public 'wasn't ready to throw out its old colonial in favor of functionalism in chromium'; in other words, modern style was too advanced for the customers' taste. And finally, 'there was a decided lack of well-designed modern furniture at anything approaching a popular price'. It was hoped, however, that at long last, 'good modern furniture, of acceptable design, is about to crash the great markets in the lower price brackets'. Already, Sears, Roebuck and Company, the great mass-merchandising firm, was featuring 'modern' furniture in its catalogue.[36]

John F. Pile, in a detailed and useful description of modern furniture making, asserts that 'a key idea of the modern movement is that the scientific and technological revolutions of the modern world have made materials and techniques available that are often better than their traditional counterparts'.[37] His description, in 1979, of how furniture is actually made, however, demonstrates conclusively that modern design and materials have not yet led to the ideal of mass production established by Ford for the automobile industry. He blames the old problem of the large number of different products that must be made, and notes ironically that the availability of good, modern design appears merely to reinforce the old pattern of smaller runs, and therefore more handwork, at the top of the scale, and more nearly 'mass' production at the lower end of the cost scale.[38]

This survey of the history of production in the American furniture industry reveals, I believe, the following characteristics. First, the industry never abandoned its primary commitment to providing the many products and styles demanded by an ever-changing consumer market. Second, within this basic commitment, the industry transferred to itself those new materials, production machines and methods from other industries which were compatible with its wide variety of product and small production runs. Third, it maintained a hiring of and respect for skilled craftspeople even while adopting various machines which would allow it to 'save labour'.

As Scranton has pointed out, the custom/batch producers in American industry have never been simply failures, frozen in an obsolete and half-realized rationalization. They have thrived, in part because of their flexibility, and have been intimately involved in the industrial evolution of the nation. This type of industry, Scranton asserts,

> shaped the image of American consumption. In a sense, batch capabilities lay behind aspects of national culture that resonated with constant change, suspected standardization and uniformity, and sustained the ideal of individualism expressed through purchase and display.[39]

It suggests that the technological history of the American furniture industry deserves to be closely studied, not for its failures of obsolescence, but for a better understanding of its successful flexibility.

On closer inspection, the demonstrable fact that the furniture industry has never found its Henry Ford appears to be only partially correct and perhaps a good thing. As segments of the manufacturing world, led by Toyota in the field of automobile production, begin to turn away from mass production toward what is sometimes called 'lean' production, the mix of production techniques which has characterized the American furniture industry during the past century now appears in a different light.[40] Indeed, the organization of machinery, labour and inventories, and the sensitivity to consumer tastes and demands which make up lean production, can all be found previously articulated in the furniture industry over the past century.

Notes and References

1. An earlier version of this paper was delivered at a conference sponsored by the Japan Science Foundation, and included in their report *US–Japan Comparison in National Formation and Transformation of Technology: Centering around Mass Production Systems, 1900–1990* (Tokyo, March, 1992), 71–96.

2. David A. Hounshell, *From the American System to Mass Production, 1800–1932: The Development of Manufacturing Technology in the United States* (Baltimore, 1984), 151.

3. Philip Scranton, 'Diversity in Diversity: Flexible Production and American Industrialization, 1880–1930', *Business History Review*, 1991, 65: 31.

4. J.D. Wallace and Margaret S. Wallace, 'From the Master Cabinet Makers to Woodworking Machinery', *A.S.M.E. Mechanical Engineering*, 1929, 51: 840.

5. The best source for this and related machine tools is Carolyn C. Cooper, *Shaping Invention: Thomas Blanchard's Machinery and Patent Management in Nineteenth-Century America* (New York, 1991).

6. Quoted in Hounshell, *op. cit.* (2), 125.

7. Quoted in James Stanford Bradshaw, 'Grand Rapids Furniture Beginnings', *Michigan History*, 1968, 52: 283–4. Unfortunately, there is no mention of the specific machines used.

8. *Ibid.*, 284.

9. *Thirteenth Annual Report of the Commissioner of Labor. 1899. Hand and Machine Labor. Volume I. Introduction and Analysis* (Washington, 1899), 269, 268.

10. *Ibid.*, 279.

11. Hounshell, *op. cit.* (2), 132.

12. *Ibid.*, 134, 142.

13. *Ibid.*, 145.

14. Scranton, *op. cit.* (3), 38.

15. Quoted in Hounshell, *op. cit.* (2), 126.

16. B.A. Parks, 'Engineering in Furniture Factories', *A.S.M.E. Mechanical Engineering*, 1921, 43: 85.

17. *Ibid.*, 86.

18. Hounshell, *op. cit.* (2), 126.

19. The following is taken from S.H. Tomlinson, 'Adjusting Manufacturing to Style Demand', *Factory and Industrial Management*, 1929, 78: 808–10.

20. H.K. Ferguson, 'Design Your Building around Your Handling', *Factory and Industrial Management*, 1928, 76: 274–6. I am grateful to Graham Hollister-Short for alerting me to G.F. Zimmer, *The Mechanical Handling and Storing of Material* ... (London, 1905), who believed that he was recording this important subject for the first time.

21. C.A. Sears, 'Lower Costs by Down-Hill Handling', *Factory and Industrial Management*, 1928, 76: 691–5.

22. M.C. Phenix, 'Line Production: We Found It the Easiest Way to Profits', *Factory and Industrial Management*, 1930, 80: 1170–1.

23. It was claimed that by the 1950s, one furniture plant, Bassett Superior in Virginia, 'produces one bedroom suite per minute'. J.L. Oliver, *The Development and Structure of the Furniture Industry* (Oxford, 1966), 117.

24. The innovation of plywood by Singer is an obvious exception. Moreover, there is some evidence that the support of upholstery borrowed innovations in the design of car seats from the auto industry.

25. Victor S. Clark, *History of the Manufactures in the United States*, III (New York, 1949 [1929]), 244.

26. In 1971, the journal *Machinery* carefully listed the various uses of metal furniture: 'bank equipment, library equipment; office equipment, which may be subdivided into the American and European systems; ship furniture; equipment for the government, such as post-office boxes, shelving and bag racks, court-house equipment, etc.; telegraph and fire-alarm equipment; interior of railroad cars and automobile bodies; factory equipment; household equipment, such as chairs, kitchen cabinets, etc.; and miscellaneous equipment, which includes articles often made of wood or other material than can be made at a profit from sheet steel.' K. George Selander, 'Production of Metal Furniture', *Machinery*, 1917, 28: 151.

27. *Ibid.*

28. *Ibid.*, 154, 152.

29. *Ibid.*, 152, 154 and 153. The James F. Lincoln Arc Welding Foundation encouraged electrical welding in industry, and in 1939 awarded a prize to a paper describing the welding of steel, tubular furniture. Clinton Bolin, 'Tubular Furniture', *Steel*, 1939, 104: 58, 63.

30. G.F. Wetzel, 'Manufacturing a Steel Chair', *American Machinist*, 1918, 48: 92–3.

31. Gilbert L. Lacher, 'Furniture – An Important New Use for Steel', *The Iron Age*, 1923, 112: 665–70.

32. The following is taken from 'Aluminum Gains Favor in the Furniture Field', *The Iron Age*, 1935, 136: 30–3. There was iron furniture, however, in the nineteenth century: see Georg Himmelfarb, *Iron Furniture* ... (Bad Ems, 1981). Also see Siegfried Giedion, *Mechanization Takes Command: A Contribution to Anonymous History* (New York, 1948), for modernist metal furniture in the twentieth century.

33. 'Aluminum Gains ... ', *op. cit.* (32), 32.

34. *Ibid.*

35. *Ibid.*, 33.

36. 'We Go Modern', *Business Week*, 1934, October: 8.

37. John F. Pile, *Modern Furniture* (New York, 1979), 147.

38. *Ibid.*, 105. On this page Pile provides a useful diagram of 'a typical factory producing a range of products in different materials'.

39. Scranton, *op. cit.* (3), 32.

40. On 'lean production', see James P. Womack *et al.*, *The Machine That Changed the World* (New York, 1990).

Space, Time and Innovation Characteristics: The Contribution of Diffusion Process Theory to the History of Technology

JENNIFER TANN

INTRODUCTION

The place of theory in the history of technology has been the subject of recent debate. This was triggered by a paper of R. A. Buchanan's[1] in which he assessed the relationship between theory and narrative. Responses were made by P. Scranton[2] and J. Law,[3] the discussion being further advanced by contributions to a session in the 1993 Oxford Conference on Technological Change entitled 'Does Theory Matter in the History of Technology?'[4] Buchanan's starting-point was that the theoretical element in the history of technology has 'received too much emphasis and the time has come to reassert the importance of narrative'. In other words what is required is not more theory but less. History, and therefore the history of technology, is, Buchanan argues, a different kind of discipline from the social sciences; when social scientists enter the field, they should observe the rules of historical procedure, for history 'should be studied as history whatever conceptual systems and attitudes are brought into it'. Buchanan reiterates the need for a first-order theory, without which a historical investigation cannot be undertaken. This comprises the original assumptions about the subject. He denies that this leads historians of technology to hanker after 'a transcendent canon of objectivity', acknowledging that first-order theory accepts that complete objectivity is unattainable. Social scientists may, however, be forgiven for wondering whether historians of technology usually make their first-order theories sufficiently explicit. It is not, however, the first-order theories but the second-order ones, those employed in the conduct of an examination of materials, that prompt Buchanan's concern. The application of second-order theories, Buchanan argues, can only be effective by 'simplifying the records by omitting facts which may be inconvenient, and by forcing the facts which are selected into patterns which may bear no relation to the circumstances of their origin. The result of such treatment ... *must* [my italics] be a distortion of the historical account.'[5]

There appears to be an underlying assumption that, unlike the 'reality' of the past, the social sciences deal with a reality which can be subjected to testing. There is not, however, a single reality and social scientists are concerned with many realities. Theories are sometimes viewed as occupying 'some rarefied space apart from the messy realm of experience, upon which they impose order'.[6] But there is no privileged place for theory and to assert that it is ideological or merely obfuscating is to fail to recognize, as Scranton reminds us, citing Heidegger,[7] that theory is a 'genetic' part of the process of being. Theory represents 'ways of looking around'. It is inescapable even to those who wish to deny its existence. Theory can become an essential part of refining the historian's enquiries and of defining the assumptions and expectations in the interpretative process. It is the identification of these different 'frames' by social scientists in their contemporary research that has prompted the application of new methodologies to the study of the history of technology. A major contribution of social science to the history of technology has been the raising of new questions through the application of theories and models.

The contribution of economics to the history of technology has been of great significance. Schumpeter focused much of his early twentieth-century work on theoretical aspects of the role and contribution of innovation to economic growth. This subsequently became a springboard for a number of distinguished contributors[8] to debates that include the economics of knowledge and invention, long and short waves, the economics of choice, production theory, the sailing-ship syndrome effect, diffusion in mature technologies, science and technology policy and counterfactual theory. While there has as yet been no integrating discussion of these literatures from the perspective of the historian of technology,[9] much has been published on specific aspects of the economics of innovation and the subject will not be addressed further in the present paper. An historiographical survey of other major strands in social science theory and their application to the history of technology shows that little attention has been paid by either historians of technology or social scientists to the contribution that innovation and diffusion process theory can make to the history of technology. In the discussion that follows some of the major strands of social science theory and its application to the history of technology will be identified as a context within which to explore the potential contribution of process theories of innovation and diffusion to the study of the history of technology; and to present a stage model of the diffusion of technology from one social system or area to another.

SOCIAL PROCESS THEORY

A narrative can be constructed in many ways and the number of different narratives that can be told about a particular phenomenon may exceed the number of identifiable facts.[10] Facts are only elements in a narrative and, while different histories can be written of, for instance, steam power and electronics, none could claim to be *the* history of steam power or electronics. Different perspectives on the history of steam power, for example, might include: government policy on the export of technology; adopters' exercise of choice in prime movers; alternatives to the Watt engine; the contribution by artisans to the development of the steam engine; intellectual property in the steam engine; overseas demand and the supply of steam power; the emergence of a new technological paradigm; a counterfactual history of the steam engine. All the above and many others might contribute to a history of steam power, and some might be written using identical sources; there is a large number of possible narratives.

The uniqueness of events is acknowledged in Bourdieu's[11] theory of practice, in which the strength of inertia in practical life is also emphasized. Bourdieu draws attention to the interplay between individuals and objective structures, between the possibilities for and the restrictions on action. Practice, for Bourdieu, is the outcome of a relationship between a situation and what he calls a habitus, a matrix of action, perception and thought. Bourdieu, in other words, acknowledges the significance of choice. And while choice, in the sense of the social determinants of technology in a specific context or situation, characterizes much of the work of sociologists of technology, a relatively ignored aspect has been an exploration of the possible alternatives available to potential adopters of technologies. Mikael Hard[12] points out the potential for a theory of practice to the history of technology in contributing to the analysis of individual and group-specific engineering, in terms of both practice and global orientations. A theory of practice, he argues, would highlight the contingent and pragmatic character of engineering, and would acknowledge the existence of what he calls inter-subjective elements, i.e. situationally specific but nevertheless comparable elements, which would not always be globally known or accepted, nor identically interpreted everywhere. Engineering in this framework can be treated as a practice that includes both abstract and concrete elements, embodied skills and abilities as well as disembodied knowledge. Technology, like science, is a practical activity that is contextually limited and only in part includes universal and cognitive elements. Technology includes knowledge components but also practical, embodied and locally delimited components. A theory of practice permits the technological historian both to acknowledge and to explore the situationally specific aspects of technology while seeking patterns within a larger framework that are only possible with reference to overarching theory.

A study that explores the particularity of technology through an ethnographic approach is B. Latour and S. Woolgrave's *Laboratory Life*,[13] in which work in an industrial research laboratory is meticulously recorded. What emerges is a richness of detail demonstrating the triviality of much scientific work, in which mundane problems are explored employing large amounts of practical skill. This approach, while intrinsically interesting, has proved to be less fruitful for the study of technology than it might have been by becoming a descriptive discourse analysis – an analysis of what people said they did, rather than what they did. Had an appropriate methodology been identified, however, the study could have been developed to explore Argyris and Schon's distinction between espoused theory and theory in use,[14] thereby turning a theoretical cul-de-sac into a study of mainstream significance for the history of technology. In the history of engineering, for example, this would permit an evaluation of the complementarity of roles of the practical mechanic and the consultant engineer at particular sites where both have left written and physical evidence; or the evaluation of a technical problem by the user of a technology compared with a third-party or *ex post facto* assessment of the situation.

P. Scranton[15] draws attention to the 'none too porous boundary' that exists between technological and labour historians. Technological historians have often allowed their fascination with people, ideas and machines to produce studies of heroes or objects which are removed from their organizational and productive contexts, while labour historians have tended to subordinate the technological to the personal and political to throw light on the workers' world. There are, for instance, relatively few articles in *Technology and Culture* and even fewer in the *Transactions of the Newcomen Society*[16] that set the machine or technological innovation in the context of implementation and use. There is, particularly in the latter journal, a tendency to a time- and context-free approach.

The labour historian has, however, much to contribute to the history of technology, although labour history has often suffered from rather primitive notions of factory practice and technical change and has tended to ignore spatial and temporal lags in diffusion. Yet the insights of labour historians into the politics of machines and systems, the gendering of technology and the cultural meaning of resistance to technology are important. Historians of technology can benefit from an exploration of how technologies were employed, appropriated and altered by workers; how new technologies were introduced into and impacted on the social relations of production. In their study of the Solihull Rover car plant, for example, R. Whipp and P. Clark[17] draw attention to the parallel production-design processes being carried out by management and labour when a new plant was being planned during a period of poor industrial relations. And Noble's[18] classic study of technological choice and appropriation in the adoption of computer numerically controlled machines by an American engineering firm highlights the contrast between intended and unintended consequences of technological innovation; management's agenda to reduce the micro-political power of operatives being reversed through up-skilling rather than de-skilling in practice.

During the 1970s sociologists and historians of the labour process focused on the work itself, the objects on which that work was performed and the instruments of work. They emphasized resistance to technology, de-skilling and the disempowerment of the labour force.[19] Later work such as that by Buraway[20] suggests alternative interpretations which emphasize consent, re-skilling or up-skilling and empowerment; examples might be the typist who becomes a skilled information processor or the boiler mechanic who becomes a nuclear technician. P.L. Frost[21] contends that by looking at sectoral political culture, economics and technologies, a new approach to understanding workplace innovation becomes possible. He suggests that the political, economic and social contexts of innovation can appear to have more impact on shopfloor power than workers' skills. Labour process historians who have rightly seen a rich vein to quarry for studies of nineteenth- and twentieth-century industry have tended to focus on manufacturing industry, in which there has traditionally been a high craft component and in which labour costs have been significant. In the process industries, however, technological innovations are more generally associated with managerial decisions located around energy efficiency, inventory control, material waste and rectification levels rather than labour costs. Only where labour has been a perceived threat to managerial power is an agenda for innovation likely to be underpinned by a desire to de-skill and disempower. It is perhaps significant that a generic technology such as steam power provoked little opposition, whereas the introduction of specific machines did. Significant though the contributions of the new social history, particularly labour process history, has been, Merritt Roe Smith[22] is right to point out the 'kaleidoscopic quality that makes their histories seem myopic' and that has moved the history of technology away from the more overarching theoretical orientations of social science, with the result that synthesis is more difficult to achieve.

SOCIAL SCIENCE THEORY AND THE HISTORY OF TECHNOLOGY

A landmark in the history of technology was the publication of Bijker, Hughes and Pinch's *The Social Construction of Technological Systems* . . . in 1987. The book is the outcome of a series of dialogues between sociologists and historians seeking to explore the interrelationships between technical, social, economic and political elements in technologies. Among a number of contributions concerned with theory in the history of innovation is Callon's,[23] in which he observes that the outcome of innovation depends on a heterogeneous network of social structures in which technologies are used. He might have also asserted that technologies shape the network. Pinch and Bijker[24] argue that technological development proceeds through variation and retention and that at any period different users or potential users may interpret an artefact differently, a point developed elsewhere by E.M. Rogers (see below). Variations in the technology will appear in relation to different perceptions of problems, while some variations will stabilize as users see particular problems solved. Hughes[25] writes of a declining innovative capacity in large organizations but, as Abernathy and Clark[26] have pointed out elsewhere, although the mature stage in a product life-cycle may be characterized by incremental innovations, major innovation may still take place in manufacturing processes. These authors and others have challenged historians of technology, particularly those in Britain, to look again at social science theories. However, one reviewer[27] of the Bijker *et al.* book suggests that the best sociologists are, in a sense, the historians who are better able to grasp the higher levels of organization in technical change. Sociologists, on the other hand, are dedicated to actor-level micro-sociological approaches.

In *The Lever of Riches*, Joel Mokyr[28] reviews the many explanations offered by earlier scholars for the occurrence of technical change in some societies and not in others. Mokyr finds most of the standard explanations for technological change, such as population growth and nutrition, war, environment and resources, labour costs and institutions, wanting. The explanations that he accepts are creativity and values that encourage it, religious beliefs that place humankind above nature, positive attitudes towards both innovation and diversity, and social values, which include a respect for manual labour, competition and business success. The barriers that he identifies include tradition, routine, vested interests, conservative elites and bureaucracies. None of these explanations would come as any surprise to scholars or practitioners of contemporary innovation, such as Rothwell and Zegveld, Gold, West and Farr or Gronhaug and Kaufmann.[29] However, they do not go far enough. Mokyr, using the analogy of biological evolution, finds equilibrium or stagnation to be the norm and innovation the issue demanding explanation. He draws attention to the clustering of micro-inventions, a phenomenon that has been extensively explored by economists such as Freeman,[30] but makes scant reference to the work of social anthropologists and psychologists, which has thrown light on the facilitating mechanisms and barriers to innovation.

THE INNOVATION AND DIFFUSION PROCESS

The relative lack of research on the history of technological diffusion may be partly explained by politics and fashions in research which have tended to promote a focus on the micro study. There has also been a move away from the use of highly quantitative models applied to the history of technology.[31] This should not, however, imply that there is a paucity of theory which could contribute to a study of the history of innovation and diffusion process. Among the questions on which light can be shed by the use of theories and models are those concerning choice, cultural imperatives, radical versus incremental change, 're-invention', the characteristics of innovations, the characteristics of innovators, stages in adoption decisions and the process of international technology transfer.

In discussing the determinants of technological choice, D. McKenzie[32] draws attention to the fact that they cannot be unidirectional. Some innovations were adopted largely on account of the apparent recalcitrance of labour, particularly skilled labour, to, for example, factory discipline. Factor prices may have been a strong determinant of choice in some situations, as, for example, where manufacturers retained the Newcomen engine in preference to the more advanced Watt steam engine.[33] In others it was the context of the evolution of relations within work-forces, and between workers and capitalists which led to different technological solutions, as, for example, the preference of Lancashire cotton-mill owners for spinning mules and New England mill owners for ring spinning.[34]

M. B. Schiffer[35] explores a model that he terms the cultural imperative. A cultural imperative is a product vision, the 'fancy of a group, often a very small group'. It is not, according to Schiffer, 'a product that everyone obviously wants'[36] (his example being the shirt-pocket radio). Here Schiffer confuses the notion of imperative, the compelling force, with the outcome, a concrete product. Cultural imperatives as compelling forces that lead to concrete products become mandates for technological development and account for the episodic and multilinear developments which are connected by networks and expectations. Early adopters of a new technology are responding to a cultural imperative in their sense of compulsion to acquire it. N. Rosenberg,[37] for example, identifies individuals who are addicted to technology, acquiring it for non-(economically) rational reasons. The historian could explore cultural imperatives in many situations. A negative example is the failure of public health advisers to introduce water boiling among High Andean peasant communities to reduce infant mortality, on account of these communities believing the boiling of water to be analogous to killing a life force.[38] Cultural imperatives probably operated alongside factory prices in the mule versus ring spinning example cited above.

The question of radical versus incremental change has caused almost as much disharmony among historians of science and technology as has that of technology push and demand pull among economists. E.W. Constant[39] suggests that the debate is less about the nature of change but rather stems from a failure to acknowledge the hierarchical structure of all complex technological systems. Systems are, he says, composed of subsystems, which are in turn composed of a wide variety of components. A system is therefore decomposable and parts can be replaced and improved, sub-problems being isolated and solved independently. Thus, whether a given change is perceived as radical or incremental depends solely on, in Constant's terms, the hierarchical level at which the issue is investigated. A new valve for an engine may represent a radical solution to a specific sub-problem but at the level of the total machine may appear only to be an incremental improvement. While Constant conceives of disaggregation in terms of level, Quinn[40] is concerned with 'logical incrementalism', by which technologies are aggregated in a horizontal plane. Quinn's major case study illustrating this concept is the Pilkington float glass process, which he views as a series of horizontally connected steps, rather than a radical innovation. But Constant and Quinn, in not accepting a scale of magnitude in innovativeness, exclude the possibility of distinguishing between, for instance, a major model change in the auto industry, such as the Rover SDI, and incremental styling innovations, which can be expected on a more or less annual basis; or the regular modifications to equipment and work organization that follow major product innovations in many industries. The integrative production engineering innovation of Richard Arkwright,[41] by which a new technology system in cotton spinning was achieved, is, in these terms, no more significant than the innovation of a single carding or spinning machine. The interconnectedness of technology is addressed by Thomas Hughes in his concept of the reverse salient, which encompasses the forward integration or knock-on effect of one technique on others, where a process ceases to be in harmony through the improvement of a connected process. Some processes may continue to be out of harmony, while for others the state is temporary. Mechanical puddling was never satisfactorily brought to a state where the reverse salient created by the bulk flows of liquid iron could be handled by this means, whereas the pressure for innovations in weaving caused by the power spinning advances of the late eighteenth and early nineteenth centuries led to effective solutions to the problem. The notion of a scale of innovativeness is implied by Hughes. By contrast, the solution of a sub-problem in Constant's terms would not impose a reverse salient.

Much of the writing on technological innovation by and for engineers[42] focuses on the manufacturer as innovator. In this approach writers are concerned with managing R&D and new product developments, sourcing innovative ideas, protecting intellectual capital, project selection, financial planning for innovation, technological forecasting and organizing for innovation. While B. Twiss[43] emphasizes that the conversion of technology into the satisfaction of a customer need is of paramount importance, he also remarks that, in the case of radical innovation, the technologist may find that 'the conventional wisdom and limited horizons of the marketing manager may well lead to the wrong conclusions'. Thus incremental innovation can occur through a functionally integrated approach to project management but attempts to achieve radical innovation may require a champion to promote a project through political alliances and financial backing, besides rational argument and technological proof.

This emphasis on the manufacturer's role, the manufacturer-active paradigm (MAP), has been challenged by E. von Hippel[44] and G. Foxall.[45] Von Hippel, as a result of research into innovation in scientific instruments, showed that many ideas for new developments originated with customers who:

- perceived that an advance in instrumentation was required;
- invented the instrument;
- built a prototype;

- proved the prototype's value by applying it;
- diffused detailed information both on the value of the invention and on how the prototype could be replicated.

Foxall showed that a parallel situation occurred in the aerospace industry. This is the customer-active paradigm (CAP). In some situations the customer provides the idea and possibly the results of initial experimentation but takes no further action after transmitting the information (CAP-1), while in others the customer takes a share in the intellectual capital (CAP-2) (Figure 1). G.A. Lee, a Manchester cotton spinner, was a CAP-1 customer. He regarded his factory as a laboratory for controlled experiments on steam power, factory heating and gas lighting, sending reports and recommendations regularly to his friend James Watt Jr, whose engineering business benefited as a result.[46] While these literatures do not consider the question of whether the creative and selective user is an innovator, R. Moss Kanter[47] asserts that any individual or organization using a technology/technique in a context that is new to him or her is innovative. To consider otherwise, she argues, is to deny the risk and uncertainty and the very real threat of failure. The human side of the adoption of new technology requires skilful managing, as the physical evidence of many abandoned micro-computers demonstrates. The constructive handling of a reverse salient requires intuition and practical

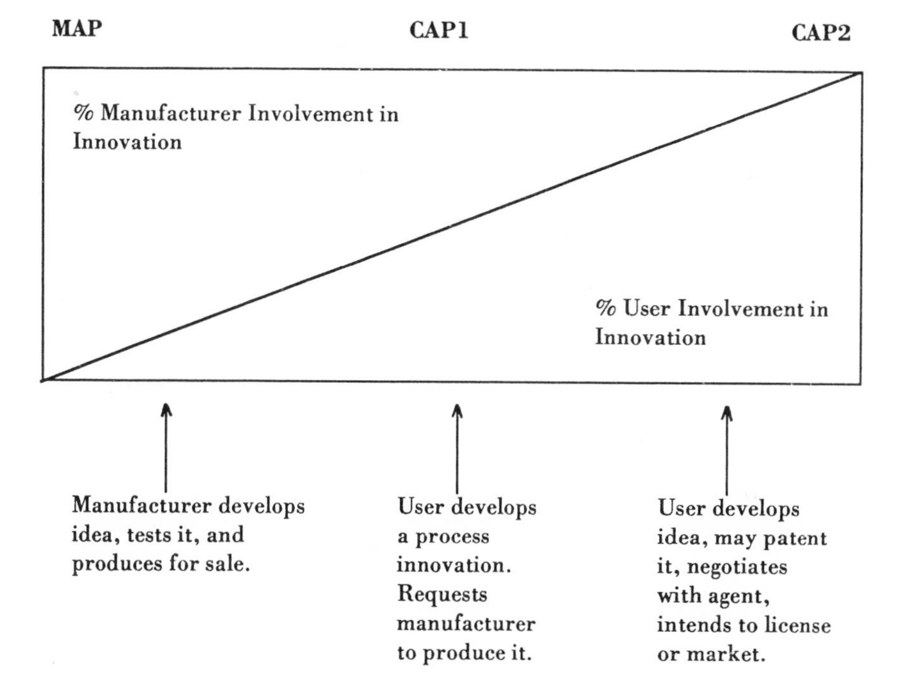

Figure 1 The manufacturer and customer – active paradigms.

skills, while creating a new system out of well-known and tried elements requires powerful alliances, effective communication and open management, besides technological skills.

Bela Gold's[48] exploration of adopters of innovations who find it difficult to characterize a particular innovation and to anticipate how its introduction might impact on existing methods of operation is relevant here. It is to be expected, he contends, that perceptions of innovations – scale, shape, uses – will vary widely between potential adopters, as will perceptions of relative advantage. To this might be added the concept of intentionality, with the important caveat that outcomes are not necessarily as intended. Innovations have come to be conceptualized as heterogeneous complexes, as bundles of elements, rather than as homogeneous entities. This was a concept anticipated by Schumpeter, employed by Hollister-Short in his longitudinal study of the 'ingredients' of the sector and chain[49] and otherwise neglected until developed by E. M. Rogers and termed by him 're-invention'.[50] The value of the concept of an innovation as a bundle of elements is that it becomes possible to map the adoption and diffusion of variants of a technological innovation more meaningfully. The concept can be characterized as follows:

- core elements and uses are distinguished from secondary ones;
- different users, even those within the same organization, perceive the configuration of the elements and their uses differently;
- the bundle can be 'unpacked', users selecting elements at will;
- users may introduce new elements;
- elements may evolve dynamically and unevenly through time (Figure 2).

The extent to which elements or bundles of elements of an innovation undergo modification, development and selection is often understated and not fully recognized in the concepts of either learning by using or learning by doing. It is an approach to technology that can be particularly helpful in the study of the international transfer of practices. The diffusion of Fordism from the United States to Britain, for example, can be more effectively understood when considered in its numerous variants at different manufacturing plants.[51] Similarly, the diffusion of computer-aided production management is more effectively understood if considered as a bundle of elements with numerous implemented variants.

In his work on the diffusion of innovations, E. M. Rogers focuses on the mechanisms of adoption and discusses the perceived characteristics of innovations which facilitate or retard their adoption. He identifies five characteristics:[52]

- relative advantage
- compatibility
- complexity
- divisibility
- observability.

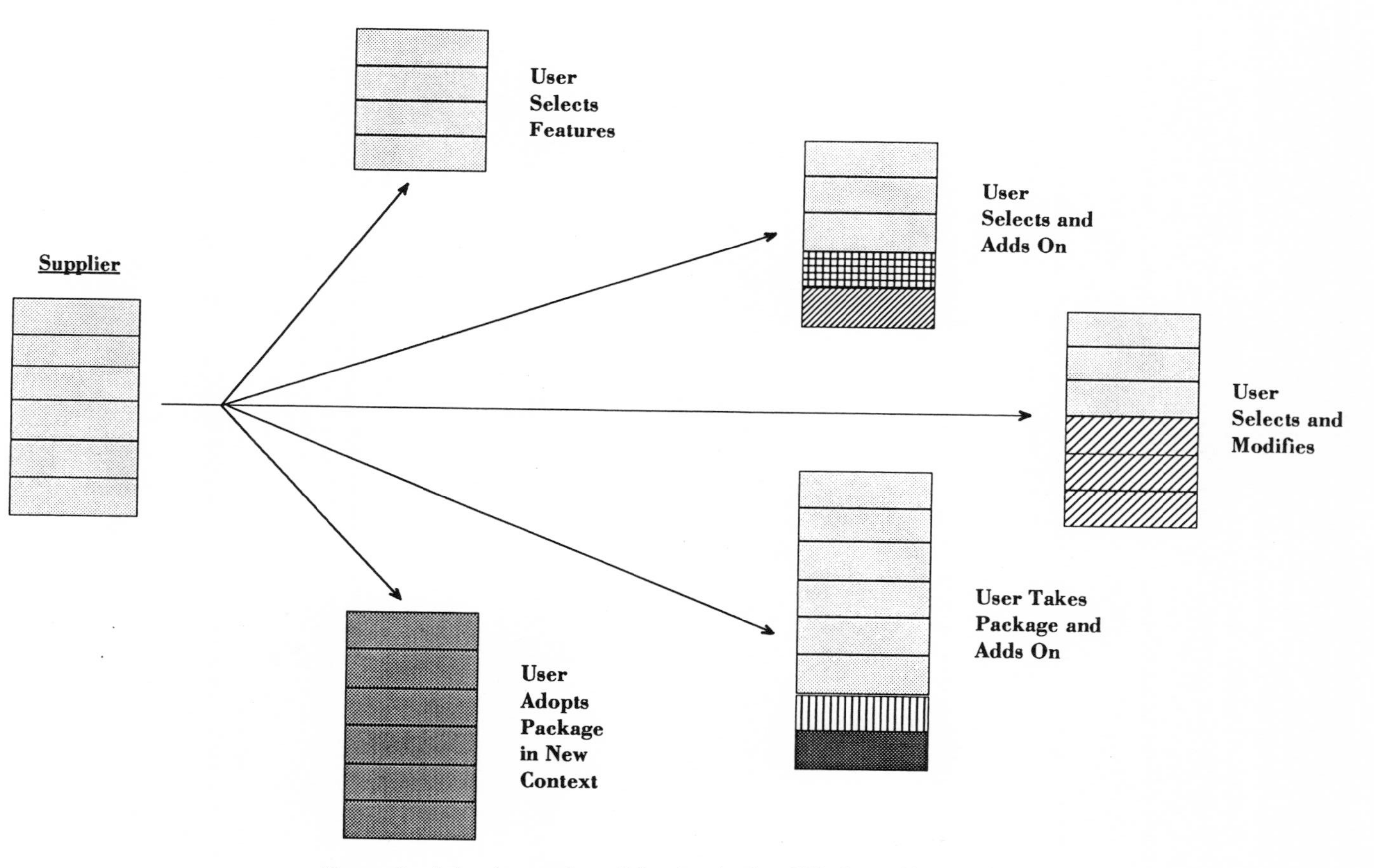

Figure 2 Selection and modification in the diffusion of innovations.

Each of these characteristics is subjective. Relative advantage will require the innovation to be perceived as materially, logically and psychologically compelling. In the diffusion of roller flour milling in Britain, for example,[53] these advantages included falling grain prices, opportunities for rapidly scaled-up production, the large-scale import of hard American wheat, which was less effectively milled by existing technologies, and a growing demand for white flour. The roller mill can be perceived as being technologically compatible with the gradual reduction (as distinct from low grinding) stone mills which had already been built in many large urban centres, rather than as a major discontinuity. There was also compatibility with regard to scale in the large gradual reduction stone plants. The extent to which a technological innovation can be understood by those responsible for making choices, installing and operating it will clearly affect the rate of adoption. At its most basic the roller mill was a simple piece of technology. But it was capable of considerable refinement and variation and the significant feature of the innovation was its development as a new milling system on the Hungarian pattern with purifiers and sifters. By the end of the nineteenth century, however, when approximately 70 per cent by volume of British-produced flour was produced in roller mills, even the most sophisticated British mill was far less sophisticated than those of Budapest. The designers of British roller mills selectively chose elements of the innovation bundle necessary to meet the pattern of demand, which was for a different quality of flour. Any technology that can be innovated on a trial basis is more likely to be adopted by a larger population of users than one that cannot. Roller mills could be introduced incrementally in combined stone roller plants, with scaling up taking place if the experiment was deemed a success. Other technologies, such as steam engines, could not be implemented as effectively on a trial basis owing to the relatively heavier costs of smaller engines, although a number of early adopters of the Watt engine followed their first orders by orders for larger engines. By comparison, the horse engine was eminently suitable for adoption on a trial basis. As a low-cost prime mover, an additional horse wheel could be added in a factory, thus taking the power to where it was needed, or, alternatively, additional animals could be attached to the original wheel.[54] An innovation whose advantages are observable is likely to have a higher rate of adoption than one where the benefits are hidden. Gas lighting in factories may not necessarily have proved an economically rational solution over oil or candles, for instance, but its observed advantages were compelling.[55]

E.M. Rogers's theoretical perspectives are rooted in social anthropology and focus on social process. His concern with the user acknowledges a greater equality of partnership between the producer/manufacturer/originator of an innovation and the adopter than do many writers on technological innovation. Rogers recognizes the risk involved for the adopter, particularly an early adopter, as his definitions of innovation and diffusion show: an innovation being something that is 'perceived as new by an individual or another unit of adoption'; and diffusion being a process by which it is communicated through certain channels in a social system over time.[56] This provides a theoretical frame of reference for those historians of technology who focus attention on the exercise of choice by users. There are four major aspects to Rogers's model. One is his focus on the antecedent context of potential adopters – their status, ability to conceptualize, values, opinion leadership and so forth, as well as the congruence of their situation with the perceived characteristics of the innovation. Secondly, he discusses the characteristics of potential adopters relative to what might be called the innovation profile for a particular innovation, relating these characteristics to economic constraints and the characteristics of the adopting unit. The characteristics of the innovation comprise the third element while the fourth is concerned with considering the innovation process in stages from knowledge to persuasion, decision, implementation and confirmation (Figure 3).[57] In Rogers's terms, the diffusion of a particular innovation among a given population is subject to the interplay between the flow of information on the one hand and the intrinsic characteristics of individuals and communities on the other. The characteristics of adopters incorporate a concept of innovativeness, which is the degree to which an individual is relatively earlier or later in adopting new ideas than other members of the social system. This concept is used to produce a taxonomy of adopter categories from innovators, early adopters, early majority, late majority, to laggards, which are presented as ideal types. The extent to which a relationship can be established between earlier or later adoption of an innovation within the life-cycle of that innovation and particular behavioural characteristics is, however, debatable. Rogers's stage model draws attention to the distinction between the diffusion of knowledge and that of practice, between know-how and how-to. It is possible, for instance, to distinguish between those individuals who were aware of Watt's engine and sought further information about it, and those who made a decision to purchase.[58] Rogers's model, while subject to the legitimate criticisms of any stage model, recognizes the possibility that an innovation decision may mean rejection rather than adoption and, while drawing attention to a post-innovation decision confirmation stage, triggers the seeking of evidence for decision-making behaviours in the history of technology. There are numerous examples from Boulton and Watt of engine customers seeking to confirm an adoption decision.[59] In some cases the expectations of the adopter were not fully met and such experiences could prompt rejection.

However, the classification of the level of innovativeness of an adopter by the stage in the cycle of innovation at which the individual adopts leads to the same measure being used both for a classification of the observed behaviour of an innovator or a laggard on the one hand, and for the stage reached in an innovation life-cycle on the other. In other words it is a quality attributed to both cause and effect. Moreover, the emphasis on adoption may lead to a pro-innovation bias. By focusing on the point of adoption rather than the point of supply or development, Rogers ignores the question of the availability of the innovation and the possibility or desirability, from the supplier's perspective, of facilitating

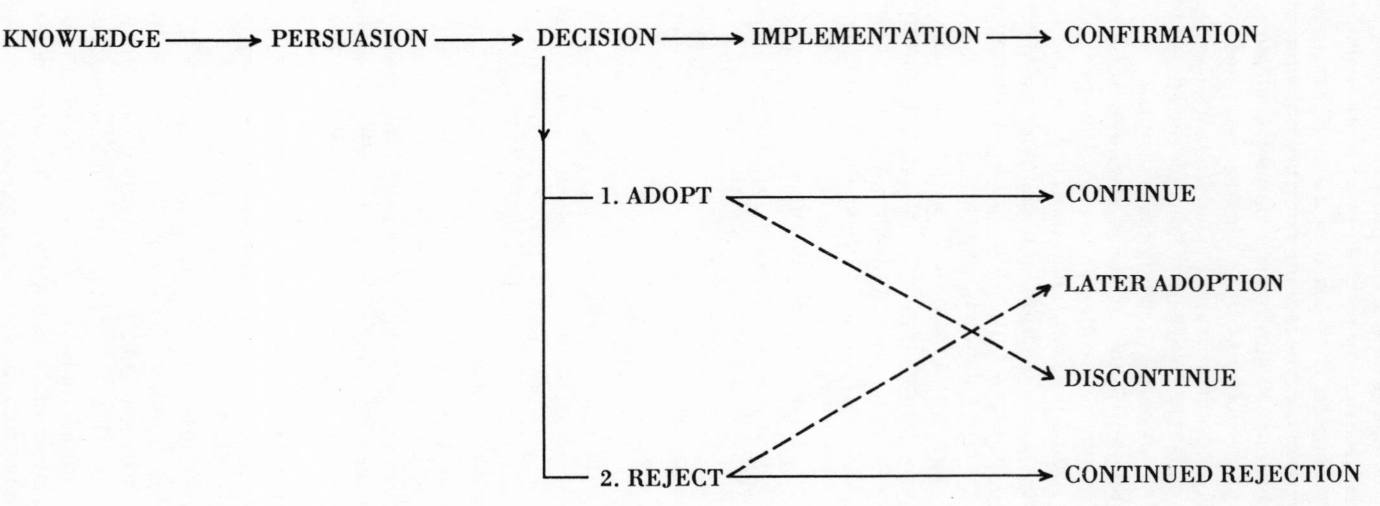

Figure 3 The innovation adoption decision
Source: Everett M. Rogers, *The Diffusion of Innovations* (1983).

the perceptual and practical disaggregation of an innovation 'bundle' into its constituent strands. Moreover, emphasis on the individual adopter, despite Rogers's assertion that this may be an organization, has limited the perceived applicability of the model, tending to exclude it from the frame where organizational process innovation is concerned. In focusing on the adoption decision to the relative exclusion of implementation, Rogers does not address the question of the subsequent adjustments made by users during or after the implementation of new technology or the difficulties experienced by users. There is extensive reliance on stylized models, such as the bell-shaped curve and the diffusion hierarchy, and a relative absence of concepts concerned with the longitudinal and embedded features of innovations and the innovation process. Innovation failure for either the manufacturer or the user is ignored not only by Rogers but by the vast majority of writers on innovation. And there is a tendency to view an innovation as being stable through time and moving at an even pace through sectors, whereas, as a study of the Watt engine shows, adoption took place unevenly through sectors over time. There is also an assumption of certainty about consequences and outcomes by Rogers, much of which is only evident *ex post facto.*

Geographers have demonstrated how diffusion tends to commence in metropolitan centres and cascade out to provincial centres and how in each locality of adoption there will be a neighbourhood effect, with the innovation following communication channels over the shortest distance in terms of time and space. Examples include the diffusion of agricultural technology such as the threshing machine and best-practice crop rotation, which tended to follow major turnpike roads or the canal network in late eighteenth- and early nineteenth-century Britain. A major problem with spatial diffusion theories (e.g. Hagerstrand)[60] is that they are stylized. There is an over-reliance on the core–periphery model, a focus on best practice rather than practices, and the implicit use of biological analogies, such as epidemics or the air-borne diffusion of seeds.

Spatial models of diffusion have not been the sole preserve of geographers, however. S. J. Mantel and G. Rosegger,[61] for example, have explored the role of third parties in the diffusion of innovations, defining third parties as persons or organizations that intervene in the adoption decisions of others. The third party may not necessarily play an independent role in the diffusion of an innovation and may have a specific interest in its diffusion. Mantel and Rosegger's typology of third parties includes: the expert decision-maker, who may or may not be a consulting engineer; a standard-setting agency, such as a professional body; an evaluating agency; a 'party at interest', which may have rights in an innovation; or an early adopter. The role of third parties in technological diffusion, termed horizontal diffusion by Rogers, is well illustrated in the diffusion of roller flour milling in Britain, with engineers fulfilling the roles of expert decision-maker and evaluating agency.

A three-stage process model developed by Tann and Breckin,[62] incorporating Mira Wilkins's concept of the absorption gap,[63] was employed in a study of the international diffusion of the Watt engine, a variant being employed by Jeremy[64] in his study of the trans-Atlantic diffusion of textile technology. This provides a macro-level framework, permitting international comparisons to be made, while elements are capable of being unpacked at different stages and locations. The model has now been revised and extended with a view to making it applicable to the diffusion of technology between social systems and geographical regions and permitting comparisons to be made (Figure 4). It is illustrated with reference to the diffusion of the stationary steam engine to the West Indies sugar plantations.

The model consists of three stages that represent different levels of intensity in the diffusion of a particular technology and its variants through a social system or geographical region. While a technology will be expected to be modified, enhanced or disaggregated, it will be deemed, for the purposes of this model, to be basically the same technology, while these changes are within the original paradigm. However, a fundamental change, as, for example, between the condensing engine and the high-pressure steam engine, will be deemed to represent a new technological paradigm. Each stage is separated from the next by a lag or gap characterized by a process of absorption of the technological concept. The imitation lag is the period during which initial awareness and knowledge of the technology are diffused within the social system or region. The subsequent absorption gaps represent periods of appropriation and embedding during which learning by using takes place and in which adopters begin to acquire a greater facility in selecting elements of the technology most appropriate to the level and location of the operations.

Stage 1 represents the initial scattering of the technology among early adopters, usually with imported technical expertise. The international diffusion of some technologies did not proceed further in certain situations, as, for example, in the diffusion of the Watt engine to India.[65] Stage 2 represents diffusion within a society or region in which there is a relatively greater number of users, and includes the employment of local technicians performing both installation and maintenance. There may be no further development from this stage, the technology in question being eventually superseded by a more advanced technology that is also imported, as, for example, in the diffusion of the steam engine to the West Indies. Stage 3 represents assimilation of the technology by the host society to the point where local manufacture commences with indigenous labour, as, for example, in the diffusion of the Watt engine to France and Germany.

The model distinguishes between relatively active and relatively passive roles for adopters between Stage 1 and absorption gap B, since it is possible for Stage 2 to be reached with a relatively high degree of reactivity on the part of adopters. A wholly turnkey operation employing expatriates would not represent assimilation.

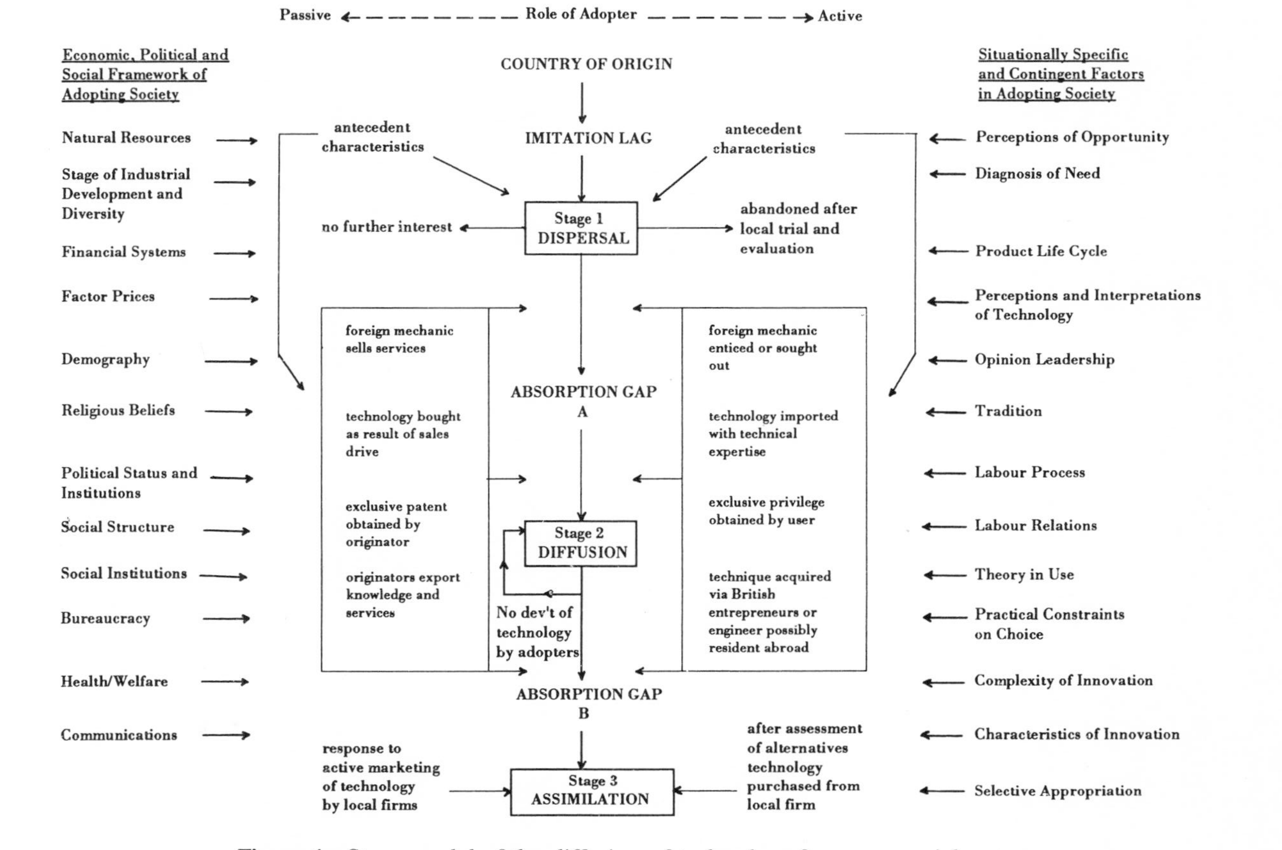

Figure 4 Stage model of the diffusion of technology from one social system or area to another.

The model is set within a contextual framework consisting of two parts: elements of the economic, political and social setting of the potential adopting social system or region; and the situationally specific and contingent factors that may influence diffusion in a potential adopting social system. These elements are derived from the overview of social science theory above, additionally incorporating elements from economics and finance.

Sales of the Boulton and Watt engine to the West Indies exceeded the aggregate of those to the rest of the world, excluding Britain and Ireland, in the 50-year period from 1775 to 1825.[66] Indeed, in volume terms engine sales to the West Indies were only exceeded in the domestic market during the same period by engines for the cotton industry (120 engines to cotton mills in Britain compared with 119 sugar-mill engines). The explanation for this uniquely intensive diffusion can be explored with reference to the essentially contingent nature of technological innovation and diffusion shown in Figure 4. In the case of the West Indies a clustering of contextual factors provided a setting unparalleled in the world at the time. These included the colonial status of the islands, the properties of the raw material to be processed and the standardized perception of need. The perceived demand was for a small, cheap and reliable engine with low running costs that could substitute either for inadequate or non-existent water-power or for cattle mills, and would be employed only intermittently. These requirements were met by 6 and 10 h.p. engines that could burn trash (dried vegetation from sugar cane). By comparison, a cattle mill required the construction of cattle and mule houses and pounds, as well as a regular supply of fresh animals.

Although enquiries were made in the late eighteenth century for engines for the West Indies, it was not until 1803 that the first order was placed for a Boulton and Watt engine, thus marking the end of the imitation lag. Stage 1 is characterized by dispersal of the first engines to different sugar-producing islands and the use of itinerant engine erectors from Britain to install them. Absorption gap A was short owing to the rapid and standardized technological solution to the provision of power for sugar mills.

One of the characteristics of the West Indies market that facilitated the diffusion of the steam engine was the integrated engineering solution to a perceived problem or need. In contrast to the supplier–customer relationship in the UK market for steam power, orders were placed after minimal technical enquiry, the reason being that the West Indies sugar-mill engine was part of a total export package of engine and mill, achieved through the collaboration of Boulton and Watt with John Rennie. Once the application of steam power to the sugar mill had been solved (and steam power had been applied to other roller processes some years before) the prototype was more or less replicated for each successive order. Thus there was a higher degree of standardization in supply terms for this market than for almost any other. Engine drawings, titled with the name of the first customer, often have up to six or more subsequent customers' names in the margin.

This standardization of supply and demand was accompanied by a proliferation of engines through the Caribbean, mostly, but not entirely, to British-owned colonies. Resident engineers installed and maintained the engines, as well as providing initial advice to potential purchasers. Stage 2 is particularly well illustrated by the West Indies. But Stage 3 was not reached owing to the absence of the necessary technical and intitutional infrastructure.

There are drawbacks to any model. In the attempt to provide a framework for longitudinal single-region studies as well as inter-societal comparisons, it has been necessary to take a holistic approach. The intention is to permit the 'what', 'when' and 'where' questions to be followed by the 'how' and 'why' ones and, it is hoped, by the 'so what' question. It is possible to lose the complexity, richness and particularity of the specific. The aim, however, is to permit data to be interrogated in such a way as not to lose narrative, while generating new questions and further levels of meaning.

It is believed that an integration of appropriate social science theory and the history of technology, with due recognition of the differing emphases that practitioners in each of the disciplines will seek, is fruitful. As technological innovators over the years have shown, it is by boundary scanning and development at the margins rather than at the core of a knowledge base that new developments take place. Whether the proposed new applications of diffusion process theories to the history of technology are radical or incremental, subsequent historians will be able to decide.

Notes and References

1. R.A. Buchanan, 'Theory and Narrative in the History of Technology', *Technology and Culture*, 1991, 32: 365–76.
2. Philip Scranton, 'Theory and Narrative in the History of Technology: Comment', *Technology and Culture*, 1991, 32: 385–93.
3. J. Law, 'Theory and Narrative in the History of Technology: Response', *Technology and Culture*, 1991, 32: 377–84.
4. R. Angus Buchanan, Mikael Hard, Philip Scranton and Hakon With Andersen, *Does Theory Matter in the History of Technology?* Workshop papers, Technological Change, a conference held at the University of Oxford, 1993.
5. R. Angus Buchanan, 'The Poverty of Theory in the History of Technology', in *op. cit.* (4), 6.
6. Paul Bove, *In the Wake of Theory* (Hanover, 1992).
7. Philip Scranton, 'Gidden's Structuration Theory and Research on Technical Change', in *op. cit.* (4), 6.
8. For example, Christopher Freeman, *The Economics of Industrial Innovation* (London, 1974); Christopher Freeman, *Design Innovation and Long Cycles* (London, 1986); Norman Clark, *The Political Economy of Science and Technology* (Oxford, 1985); Arnold Heertje, *Economics and Technical Change* (London, 1973); Stuart MacDonald, D.McL. Lamberton and T.D. Mandeville, eds, *The Trouble with Technology* (London, 1983); P.A. David, *Technical Choice, Innovation and Economic Growth* (Cambridge, 1979); Nathan Rosenberg, *Inside the Black Box ...* (Cambridge, 1982); see also Andrew Jamison, 'Technology's Theorists: Conceptions of Innovation in Relation to Science and Technology Policy', *Technology and Culture*, 1989, 30: 505–33.

9. A paper on the subject is in preparation by the author. The sailing-ship syndrome is a term used for the continued technological development and use of a superseded technology, taking its name from the improvement of the sailing ship after the introduction of the steam ship.

10. Hakon With Andersen, 'All Else is Ornament?', in *op. cit.* (4), 13.

11. Pierre Bourdieu, *The Logic of Practice* (Stanford, CA, 1990). Bourdieu's theory of practice is discussed by Mikael Hard, 'Historians as Sociologists – Technicians as Practitioners', in *op. cit.* (4), 11–15.

12. *Ibid.*

13. Bruno Latour and Steve Woolgar, *Laboratory Life: The Social Construction of Scientific Facts* (London, 1979).

14. C. Argyris and D.A. Schon, *Organisational Learning: A Theory in Action Perspective* (London, 1978).

15. Philip Scranton, 'None-too-porous Boundaries: Labor History and the History of Technology', *Technology and Culture*, 1988, 29: 722–43.

16. An exception being Michael Duffy, 'Technomorphology and the Stephenson Traction System', *Transactions of the Newcomen Society*, 1982–3, 54: 57.

17. Richard Whipp and Peter Clark, *Innovation and the Auto Industry* (London, 1986).

18. D. Noble, *America by Design: Science, Technology and the Rise of Corporate Capitalism* (New York, 1977).

19. H. Braverman, *Labor and Monopoly Capital: The Degradation of Work in the Twentieth Century* (New York, 1974).

20. Michael Burawoy, *Manufacturing Consent* (Chicago, 1979). See also Eric Batstone, Stephen Gourlay, Hugo Levie and Roy Moore, *New Technology and the Process of Labor Regulation* (New York, 1987).

21. Robert L. Frost 'Labor and Technological Innovation in French Electrical Power', *Technology and Culture*, 1988, 29: 865–87.

22. Merritt Roe Smith, 'Industry, Technology and the "Labor Question" in 19th Century America: Seeking Synthesis', *Technology and Culture*, 1991, 32: 555–70.

23. Michael Callon, 'Society in the Making: The Study of Technology as a Tool for Sociological Analysis', in Wiebe E. Bijker, Thomas P. Hughes and Trevor Pinch, eds, *The Social Construction of Technological Systems* (Cambridge, MA, 1987), 83–103.

24. Trevor F. Pinch and Wiebe E. Bijker, 'The Social Construction of Facts and Artifacts', in *op. cit.* (23), 17–50.

25. Thomas P. Hughes, 'The Evolution of Large Technological Systems', in *op. cit.* (23), 51–82.

26. W.J. Abernathy and K.B. Clark, *Industrial Renaissance: Producing a Competitive Future for America* (Boston, 1983).

27. Susan E. Cozzens, 'Review of The Social Construction of Technological Systems', *Technology and Culture*, 1989, 30: 705–7.

28. Joel Mokyr, *The Lever of Riches* (New York, 1990).

29. Roy Rothwell and W. Zegveld, *Innovation and the Small and Medium Sized Firm* (London, 1982); Bela Gold, ed., *Technological Change: Economics, Management and Environment* (Oxford, 1975); Michael West and James L. Farr, *Innovation and Creativity at Work* (Chichester, 1990); Kjell Gronhaug and Geir Kaufmann, *Innovation: A Cross-disciplinary Perspective* (Oslo, 1988).

30. Christopher Freeman, John Clark and Luc Soete, *Unemployment and Technical Innovation* (London, 1982).

31. G.N. von Tunzelmann, *Steam Power and British Industrialisation to 1860* (Oxford, 1970).

32. Donald Mackenzie, 'Marx and the Machine', *Technology and Culture*, 1984, 25: 473–502.

33. Jennifer Tann, 'Boulton and Watt's Organisation of Steam Engine Production Before the Opening of Soho Foundry', *Transactions of the Newcomen Society*, 1977–8, 49: 41–56.

34. William H. Lazonick, 'Production Relations, Labor Productivity, and Choice of Technique: British and US Cotton Spinning', *Journal of Economic History*, 1981, 41: 491–516. See also Lars G. Sandberrg, 'American Rings and British Mules: The Role of Economic Rationality', *Quarterly Journal of Economics*, 1979, 3: 213–62.

35. Michael Brian Schiffer, 'Cultural Imperatives and Product Development: The Case of the Shirt-pocket Radio', *Technology and Culture*, 1993, 34: 98–113.
36. *Ibid.*, 99.
37. Nathan Rosenberg, *Inside the Black Box: Technology and Economics* (Cambridge, 1982).
38. Everett M. Rogers, *Diffusion of Innovations* (New York, 1983), 1–2.
39. Edward W. Constant II, 'The Social Locus of Technological Practice: Community, System, or Organisation?', in *op. cit.* (23), 232–42.
40. B. Quinn, *Strategy for Change: Logical Incrementation* (1980).
41. Jennifer Tann, 'Richard Arkwright and Technology', *History*, 1973, 58: 29–44.
42. See journals such as *Technovation.*
43. Brian Twiss, *Managing Technological Innovation*, 4th edn (London, 1992), 4–5. See also E. Rhodes and D. Wield, eds, *Implementing New Technologies* (Oxford, 1985).
44. Eric von Hippel, *The Sources of Innovation* (New York, 1988).
45. Gordon Foxall and Janet Tierney, 'From CAP1 to CAP2', *Management Decision*, 1984, 22: 3–15.
46. Jennifer Tann, *The Development of the Factory* (London, 1971).
47. Rosabeth Moss Kanter, *The Change Masters* (London, 1984).
48. Bela Gold, 'On the Adoption of Technological Innovations in Industry: Superficial Models and Complex Decision Process', in Stuart MacDonald, D.McL. Lamberton, and T.D. Manderville, eds, *The Trouble with Technology* (London, 1983), 104–21.
49. G. Hollister-Short, 'The Sector and Chain: An Historical Enquiry', *History of Technology*, 1979, 4: 149–85. J.A. Schumpeter, *Business Cycles: A Theoretical, Historical and Statistical Analysis of the Capitalist Process* (New York, 1939).
50. Rogers, *op. cit.* (38).
51. Jennifer Tann, 'Diffusion of Management Thought and Practice, 1880–1970', in David J. Jeremy, ed., *The Transfer of International Technology – Europe, Japan and the USA in the Twentieth Century* (Aldershot, 1992), 193– 220.
52. Rogers, *op. cit.* (38), 210–38.
53. Jennifer Tann and R. Glyn Jones, 'Technology and Transformation: The Diffusion of the Roller Mill in the British Flour Milling Industry 1870–1907', *Technology and Culture*, 1996 (forthcoming).
54. Tann, *op. cit.* (46).
55. *Ibid.*
56. Rogers, *op. cit.* (38), xviii,5. See also Peter Clark and Neil Staunton, *Innovation in Technology and Organisation* (London, 1989), 124–39.
57. *Ibid.*, 163–209.
58. Jennifer Tann and M.J. Breckin, 'The International Diffusion of the Watt Engine', *The Economic History Review*, 2nd series, 1978, 31: 541–64.
59. Tann, *op. cit.* (33).
60. T. Hagerstrand, *Propagation of Innovation Waves* (Stockholm, 1983).
61. S.J. Mantel and G. Rosegger, 'The Role of Third Parties in the Diffusion of Innovations: A Survey', in Roy Rothwell and John Bessant, eds, *Innovation, Adaptation and Growth* (Amsterdam, 1987).
62. Tann and Breckin, *op. cit.* (58).
63. Mira Wilkins, 'The Role of Private Business in the International Diffusion of Technology', *Journal of Economic History*, 1974, 34.
64. David J. Jeremy, *Transatlantic Revolution* (Oxford, 1981).
65. Jennifer Tann and John Aitken, 'The Diffusion of the Stationary Steam Engine from Britain to India 1790–1830', *The Indian Economic and Social History Review*, 1992, 29: 199–214.
66. Jennifer Tann, 'The Diffusion of the Stationary Steam Engine to the West Indies Sugar Plantations' (forthcoming).

The German Miners at Keswick and the Question of Bismuth

DAVID BRIDGE

ABSTRACT

The German account books for the Elizabethan copper industry at Keswick in Cumberland record that by early 1569 a small quantity of bismuth had been extracted. An investigation is made into the source of this unexpected product in the light of known occurrences of bismuth in the area, impurities in the local copper ores and the smelters' understanding of them, and the efforts of one particular miner from the Tyrol. It is hoped that this work throws a little more light on the extent of the activities of the Company of Mines Royal in the north of England.

INTRODUCTION

Several attempts were made by Henry VIII to develop the mining of copper and precious metals in England but it was not until the reign of Queen Elizabeth that with the help of German expertise the Keswick copper industry finally got under way. In the early years of the enterprise (i.e. from 1564 to 1577) the Augsburg business house of David Haug, Hans Langnauer and Co. provided a large part of the financial backing and also the skills and technical knowledge needed to set up the industry.[1] Daniel Höchstetter, an associate of the company at that time, managed the works at Keswick and also held shares on Haug and Co.'s behalf in the controlling partnership of English and German capitalists, which in 1568 became known as the Company of Mines Royal.[2] Early this century the German account books for that period were discovered in the Augsburg archives by Collingwood, who took on the task of transcribing and translating the contents.[3] Although few records remained for the first years of operation, detailed Keswick 'Journals' were found to exist for the years 1569, 1571 and 1573 to 1577, and the Germans also left accounts relating to their business in London. Without doubt these meticulously kept records provide a unique insight into the Keswick mining scene, but because of the missing years and the fact that they were only intended to record income and expenditure, many questions regarding the extent of the mining operations at the time and the location of some of the smaller workings remain unanswered. This is not helped by the unorthodox spelling of English names, often barely recognizable, as the accountant struggled with a strange language and dialect.

The accounts list term by term the expenses of raising, dressing and carting copper ore from mines in Newlands and Borrowdale and also from smaller ventures such as one at Buttermere. Copper ore raised at Caldbeck is also covered but the important reserves at Coniston had not at that time been exploited. Also recorded are the consignments of bulk copper supplied to the Crown, which were shipped from Newcastle to London and delivered to the Lieutenant of the Queen's Ordnance at the Tower. From 1572, when export restrictions on copper had been lifted, foreign sales begin to feature, e.g. in Bordeaux and Muscovy. In July 1574 the first coppersmiths arrived in Keswick and soon a large water-powered copper-hammer was under construction. During the next three years, after a permit had been obtained from the Mineral and Battery Works, which held the monopoly for brass making and battery, sales of wrought copper in the form of utensils and copper sheet figure both locally and in London. Lead mining is also recorded, mainly at Caldbeck, Grasmere and 'St Joseph's' at Newlands, and before the year 1572 additional lead ore was brought in from other areas, including Teesdale and Allendale, as well as lead from Richmond and, at the start of operations, from the Bleiberg mines of Carinthia. This was needed for the extraction of silver from the copper and the lead ore provided an additional (and indeed the principal) source of silver. The company required lead for building purposes and eventually found a market for their excess lead. The accounts record silver separation by Christmas 1569, though apparently it was not until August 1572 that a sample was sent to London. By 1574 regular deliveries of silver were being made to Richard Martin, goldsmith and Master of the Mint.

In early 1569 an unexpected product appeared on the scene. On 31 January of that year 14*s.* 6*d.* was paid to 'Old Stable', a partner of James and William Stable, the London carriers employed by the company, for the carriage of four hemispheres of bismuth weighing 147 pounds. These were sent in a cask from Keswick to London and Antwerp. At the same time 1*s.* 1*d.* was paid for transporting another cask of bismuth to London. This is the only mention of bismuth in Collingwood's transcript of the accounts.

EARLY REFERENCES TO BISMUTH

In the early days bismuth was frequently confused with other metals with which it was associated such as lead, antimony and tin, and it was not until the middle of the eighteenth century that the independent researches of the German chemist Johann Heinrich Pott and the Frenchman Geoffroy established beyond doubt that bismuth was a specific element.[4] Nevertheless, a hundred years earlier the Spanish priest and geologist Alonzo

Alvaro Barba clearly recognized the uniqueness of the metal 'discovered a few years ago in the Sudnos mountains of Bohemia' in his attack on the belief held by astrologers that each of the seven metals of antiquity (gold, silver, copper, iron, lead, tin and quicksilver) was influenced by one of the seven planets,[5] and as early as 1530 Georgius Agricola had spoken of bismuth in *Bermannus*, his first work relating to mining.[6] He coined the latin name '*bisemutum*' and described the metal as '*plumbum cinereum*' or ash-coloured lead, to differentiate it from '*plumbum nigrum*' or true lead, adding that 'it generally indicates that there is silver beneath the place where it is found, and because of this our miners are accustomed to call it the roof of silver'. The earliest reliable reference to bismuth appears to have been in about the year 1500 when '*wismuth*' was mentioned in association with silver in a small book on mining geology called *Eyn Nützlich Bergbüchlein*, thought to have been written by Dr Ulrich Rühlein von Kalbe, burgomaster at Freiberg in Saxony, who helped to survey and design the mining towns of Annaberg and Marienberg.[7]

There is little doubt that the Germans at Keswick knew what metal they were dealing with. In his *De re metallica*, published in 1556, Agricola described seven different practices for extracting bismuth from its ores, most of them variations on the simple method of heating the ore on an open wood fire in a natural draught and collecting the bismuth that dripped out of the bottom.[8] This method would be suitable for native bismuth, which can be extracted by simple liquation (its melting-point being only 271° C), but in one case the ore was reduced with charcoal in a furnace 'similar to the iron furnace', suggesting that secondary ores, i.e. the oxide bismite and possibly the carbonate bismutite, were also being worked at that time. The above three ores, particularly native bismuth, were found associated with cobalt or silver in the mines of Saxony and Bohemia, notably at Schneeberg and Joachimsthal (now Jáchymov) in the Erzgebirge mountains on the border between what was until recently East Germany and Czechoslovakia,[9] a mining area familiar to Agricola, who was born in that region and for a time held the post of physician at Joachimsthal. The sulphide ore bismuthinite, which also occurs in the area (and is the principal bismuth ore found in the Lake District), would probably need fusing with iron to remove the sulphur,[10] a procedure not described by Agricola, but it is significant that in the methods described the extracted bismuth was poured into, or re-melted in, round-bottomed iron pans from which it was later removed in the form of hemispherically shaped cakes such as those described in the Keswick accounts (Figure 1).

LOCAL OCCURRENCE OF BISMUTH

This then prompts the question: where did the Keswick bismuth come from? Of the three sites in the Lake District traditionally noted for the occurrence of bismuth,[11] the deeper levels of Coniston copper mine and Shap granite quarry can be ruled out immediately. The third site is Carrock mine in the Caldbeck Fells, where bismuth has been found in

A—PIT ACROSS WHICH WOOD IS PLACED. B—FOREHEARTH. C—LADLE. D—IRON MOULD. E—CAKES. F—EMPTY POT LINED WITH STONES IN LAYERS. G—TROUGHS. H—PITS DUG AT THE FOOT OF THE TROUGHS. I—SMALL WOOD LAID OVER THE TROUGHS. K—WIND.

Figure 1 Engraving from *De re metallica* showing extraction of bismuth.

the north–south tungsten-bearing veins, but there is no direct documentary evidence that these veins were worked before the year 1854.

Up to 1569 the copper ores smelted at Keswick came principally from two sources: Goldscope mine (and possibly other lesser veins in the Newlands valley) and the Copper Plate vein in Borrowdale.[12] It has long been recognized that small quantities of bismuth often occur with copper sulphide deposits[13] – in fact one commercial source of bismuth today is a by-product from the treatment of copper ores.[14] When interest in copper mining was revived in the Duke of Somerset's Cumberland estates after the civil wars the mineral collector Dr John Woodward suspected the presence of bismuth in certain specimens of grey ore taken from Goldscope mine and St Thomas' Work in the Newlands valley.[15] Woodward was possibly the first person in Britain to collect geological specimens for their scientific interest rather than as mere curiosities, and began his well-documented collection, which is still to be found in the Sedgwick Museum at Cambridge University, in about 1696.[16] He classed these grey ores as 'marcasites', which, he claimed, 'all hold copper more or less', but his specimen descriptions suggest that they were either of the tennantite–tetrahedrite series (i.e. mixed sulphides of copper, iron and either arsenic or antimony, which are of variable composition[17] and known sometimes to contain bismuth[18]) or predominately arsenopyrite – minerals which have since been recorded at Newlands.[19] It was 'grey ores' such as these that Thomas Robinson, Rector of Ousby, found predominant when he first surveyed Goldscope in 1697 with a view to re-working the mine for copper, ores for which he never managed to obtain a satisfactory analysis even from the leading experts of the day.[20] Presumably he would only have inspected the workings above adit, where the principal copper ore, chalcopyrite, had already been worked out by the Germans.

One cannot attach too much weight to Woodward's observations, which lacked analytical confirmation, but recently bismuth has been positively identified in the Newlands copper veins by Stanley and Vaughan[21] from micro-analysis of the ores. In the Dale Head North vein (otherwise known as Long Work), small concentrations of bismuthinite in the form of aggregates up to 0.2 mm were found in association with other sulphides, together with lesser amounts of native bismuth. Similar mineral assemblages were found at Goldscope, in small copper veins at Castle Nook (a group including St Thomas' Work) and in the Copper Plate vein of Borrowdale. Even if the German miners had ignored the grey ores these recent studies leave little doubt that a small quantity of bismuth would have found its way to the Keswick smelthouses with the chalcopyrite, or *ganz erzt* as the Germans called it.

As for the lead ores, a Lake District survey of mineralization also carried out by Stanley and Vaughan[22] makes no reference at all to bismuth in the lead-zinc veins and, according to Dunham, apart from an isolated occurrence in Weardale, only trace levels have been found in galena throughout the whole of the Northern Pennine orefield.[23]

SMELTING OF THE COPPER ORES

Methods developed in the Keswick smelthouses to cope with the different types and grades of ore from the different mine workings were recorded in a notebook by Daniel Höchstetter (junior) from an accumulation of notes and papers going back half a century. This collection was started in 1615 and as well as giving detailed instructions for mixing and smelting the ores to produce malleable copper and for the extraction of silver it includes prescriptions for dry assays and also accounts and other affairs up to the year 1639. The notebook eventually came into the hands of the Duke of Northumberland and has recently appeared in print together with a second notebook containing drafts and copies of letters written by Daniel (junior), a total of 324 pages of manuscript.[24]

These notes show that the extraction of copper from the sulphide ores required a long series of roastings and smeltings, which could involve as many as 20 firings and take up to 6 months. Essentially the method consisted of a preliminary roast at low temperature to burn off some of the sulphur, after which the ore was smelted at a moderate temperature with a limestone flux to remove the gangue material and produce a matte or regulus, which they called 'green stone'. A series of roastings at progressively higher temperatures followed to oxidize and remove most of the remaining sulphur to produce 'black copper', which was then smelted and reduced to 'rough copper'. Finally, this was refined by smelting with charcoal to obtain the ductility necessary for the manufacture of battery ware. When silver was to be extracted, further stages were introduced involving the addition of lead or lead ore (i.e. 'drowning the copper with lead') and then separating the lead from the copper. During this process the silver concentrated in the lead-rich layer and was afterwards extracted from the lead by a well-established cupellation method.

Any native bismuth present would run out during the early stages and it is expected that most of the bismuth in sulphide form would follow the copper into the matte and separate out at a later stage, a fraction perhaps fuming off.[25] During silver extraction some of the bismuth may have gone into solution with the lead but at least part of that from the original ore should have been recoverable from the bottom of the furnace. Yet there is no evidence from the Höchstetter notebooks or from any other published correspondence relating to the industry[26] that the Germans at Keswick recognized the presence of bismuth while smelting copper ores.

Apart from selling off excess lead the operators appear to have shown little or no interest in by-products from their regular smelting runs before 1579. By that time the shareholders had become disillusioned. Despite the additional income from sales of battery they were continually being asked to provide more capital owing to the depressed state of the copper market,[27] and to make matters worse Haug and Co. had been taken over by their creditors and had pulled out.[28] At the end of 1575 Queen Elizabeth had advanced a loan of £2,500 and a request for a further loan seems to have fallen on stony ground. So in May 1579 the shareholders agreed to provide £100 for trials on a new method for recovering sulphur for the manufacture of gunpowder (presumably for ordnance) invented by the chemist Henry Pope.[29] They were also showing increasing concern over Daniel Höchstetter's handling of the business and were encouraging independent experts to try out the ores and inspect the mines, hoping for a better deal than they were getting out of Höchstetter. Consequently, in April 1579 two German mining experts, 'Hugh Brinckhurste the younger (of Erffurt) and Phillipp Bayer of Strassburg', came to Keswick and from an assay on the copper ores claimed that three times as much copper could be extracted.[30] But that exercise came to naught and Daniel Höchstetter, understandably annoyed by 'diverse slanderous reports' and 'secret dealings' behind his back,[31] proposed that he should take on the burden of the running costs himself and that the mines should be leased to him in return for a fixed rent. In the event the company granted a lease to Daniel Höchstetter and Thomas Smythe, the latter a wealthy merchant and Collector of Customs at the Port of London,[32] and in 1581, the year of old Daniel's death, independent advice on the smelting methods was again sought. In that year George Nedham, a company shareholder who had been actively involved in the industry, brought a Jewish chemist and metallurgist to Keswick. His name was Joachim Gaunse.[33]

Gaunse (Gans or Ganz) made a careful study of the copper ores and after some experiments of his own managed to identify 'tenn severall substances'. One of these was the copper itself and the other nine, described by Nedham as 'hurtfull and venomous humors to the copper', were identified as 'Iron, a kinde of black stone (wherin the copper groweth), a kinde of white stone named sparr, sulpher, arsenique, antimony, vitriall, calcator [calc-ochre or iron oxide], and allom.' Not only was Gaunse able to recommend a drastic reduction in the number of firings needed to reach the rough copper stage, but in doing so he was able to present the company with two commercial by-products, vitriol and copperas, which he dissolved out by 'letting water passe through the ures' after the first roast. The latter was in great demand for dyeing cloth and, although Nedham claimed to have some knowledge of vitriol production (presumably meaning blue vitriol or copper sulphate), the secret of separating the iron from the copper in sulphate form was something new to Keswick. The potential market for copperas he extolled at length: 'For vent of this Coppris ther wilbe great quantitie used in Cumberland, Westmorelande, Yorkshire, Cheshire, and Lancashire onely for dyeing ... and likewise ther wilbe much soulde in the northpartes of Scotlande' – areas previously forced to import the product from foreign countries. But nowhere in the reporting of Gaunse's work (two long documents in the Domestic State Papers[34]) or in any other development of by-products at this time do we find a mention of bismuth. Moreover, Gaunse commented on the ignorance of the German operators when it came to understanding the impurities in their copper ores.

A MINER CALLED FECHTENBACH

All the evidence therefore points to the bismuth consignment of 31 January 1569 being the result of a trial run on a bismuth-rich ore not normally encountered at Keswick – and here Bernhart Fechtenbach may provide the clue. Fechtenbach was one of the earliest miners to arrive at Keswick from the Schwaz area of the Tyrol and according to the accounts was employed at different times both as a pickman and in the Keswick smelthouses.[35] In the early days of the industry a considerable amount of prospecting must have been carried out, as shown by a letter written in 1568 by Daniel Ulstat, deputy governor of the Company of Mines Royal, which states 'We have also in that short time, since I did write, here found eighteen new mines, the which according to the assay made therof might be wrought with good profit.'[36] Fechtenbach also made a contribution in this field. In March 1574 he was paid 10*s.* for discovering a new mine near Wythburn, where he subsequently raised 22 kibbles of ore, and at an earlier date, i.e. late 1568, we find him working alone on a contract developing the '*Waiss nit*' vein called 'Windenburg', the exact location of which is not recorded, though it appears under the Newlands accounts. It is also noted that at some time in early 1569, possibly 12 January (the date is not clear), 6 kibbles of unspecified ore were carried from Fechtenbach's Nick in 'Brandlgil' at 6*d.* per kibble.

Collingwood identifies Brandlgil as one of the water courses down Cat Bells into Brandlehow Wood. A hand-chiselled adit level proves that the lead-bearing Brandlehow lode was worked in the sixteenth or seventeenth century and there is documentary evidence that there were workings known as 'Minersputt' here as early as 1566.[37] Furthermore, from Höchstetter's notes we learn that Brandlehow *ganz* ore was used in trials for the extraction of silver in 1567.[38] But in the case of Fechtenbach's Nick we happen to know the name of the ore carrier, a Richard Hodgson of Threlkeld who was largely employed in transporting peat from mosses to the north and east of Keswick:

Early 1569	Richardt Hudgson, carrying peat from Skiddaw.
	Richardt Hudson, carrying peat from Skiddaw.
	Richardt Hogdson, carrying ore from Fechtenbach's Nick.
Summer 1569	Richardt Hudgson, carrying peat from Flasco.
1571	Richard Hudson of Trelkhet, carrying peat from Skiddaw and Flasco.
1574	Ritzart Hütson of Drilket, creditor.

Why should a Threlkeld carrier with a team of workers cutting, drying and stacking peat on Skiddaw and carting it to the smelthouses in early 1569 take on an isolated job at Brandlehow across the River Derwent when there were other carriers in that area? Moreover, why should the company be prepared to pay 6*d.* per kibble for ore to be transported only a mile along the lake shore from Brandlehow to its landing stage at

Copperheap Bay in the 'Vorwald', from where it would be taken by boat to Keswick, when the going rate for carting ore the greater distance from God's Gift (Goldscope) to the landing stage was only about 1*d.* per kibble?

Now if by Brandlgil the accountant meant Brandy Gill, situated eight miles north of Keswick in the Caldbeck Fells, and included the earlier entry with the Newlands returns by mistake, then the employment of a Threlkeld carrier at a rate of 6*d.* per kibble would make sense (see Figure 2). And there is supporting evidence for this. In 1747 an anonymous traveller journeyed to the Lake District to visit the graphite mines in Borrowdale but, finding them closed, diverted his attention to the Caldbeck Fells. His description of the visit appears in the *Gentleman's Magazine* together with a crudely drawn pictorial map of the area[39]

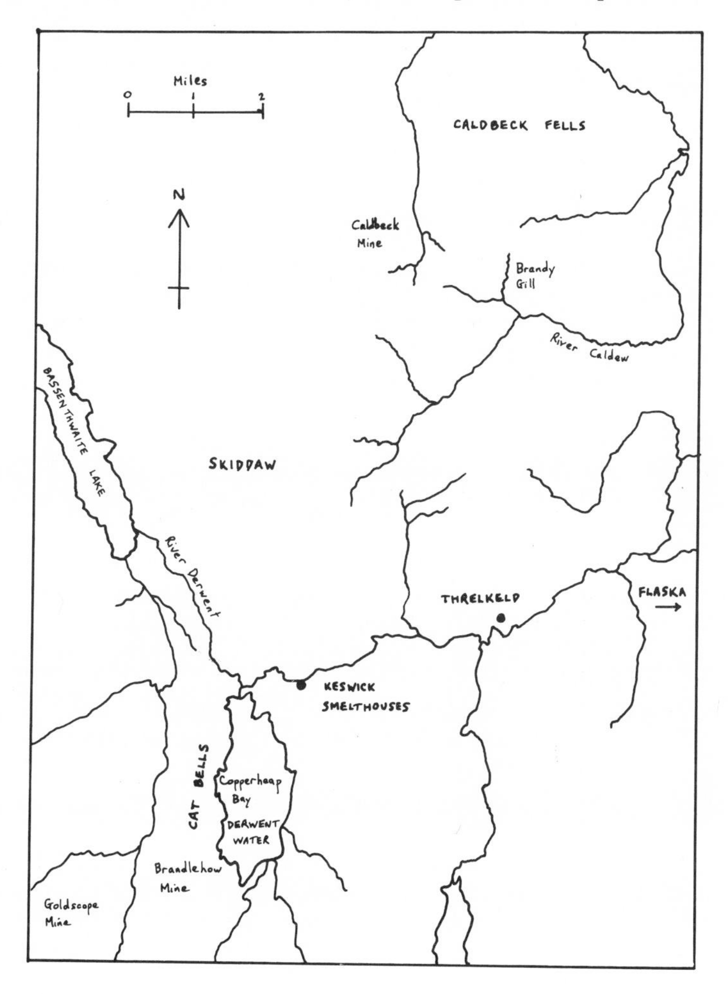

Figure 2 Mine sites in the Keswick area.

showing a gill to the north of the River Caldew below Carrock Heights, to which he assigns the name 'Brandle-gill Beck' (Figure 3).

This then raises the question: what interest would the German miners have had in Brandy Gill? There are two east–west lead-bearing veins, the upper of which is said to have also yielded copper.[40] This was developed in the 1720s and again in Victorian times from levels on both sides of the gill (see Figure 4).[41] The anonymous visitor of 1747 spoke of abandoned copper mines there 'long since worn out'. It is also recorded that lead or copper mining began at or near Brandy Gill as early as the sixteenth century.[42] But the evidence here suggests that in late 1568 the Germans, perhaps Fechtenbach himself, had hit upon one of the strong north–south tungsten-bearing veins. Could it be that the Windenburg vein (which Fechtenbach was working in the winter and spring of 1568–9) and Fechtenbach's Nick (from which ore was carted in early 1569) both referred to the same vein, Emerson's or Harding's, and that the nickname '*Waiss nit*' or 'Don't know' described its chief ore, wolfram, which they were unable to identify? In their search for copper and lead could they have recognized the bismuth ores and, knowing that in the mines of Saxony and Bohemia bismuth generally indicated that there was silver

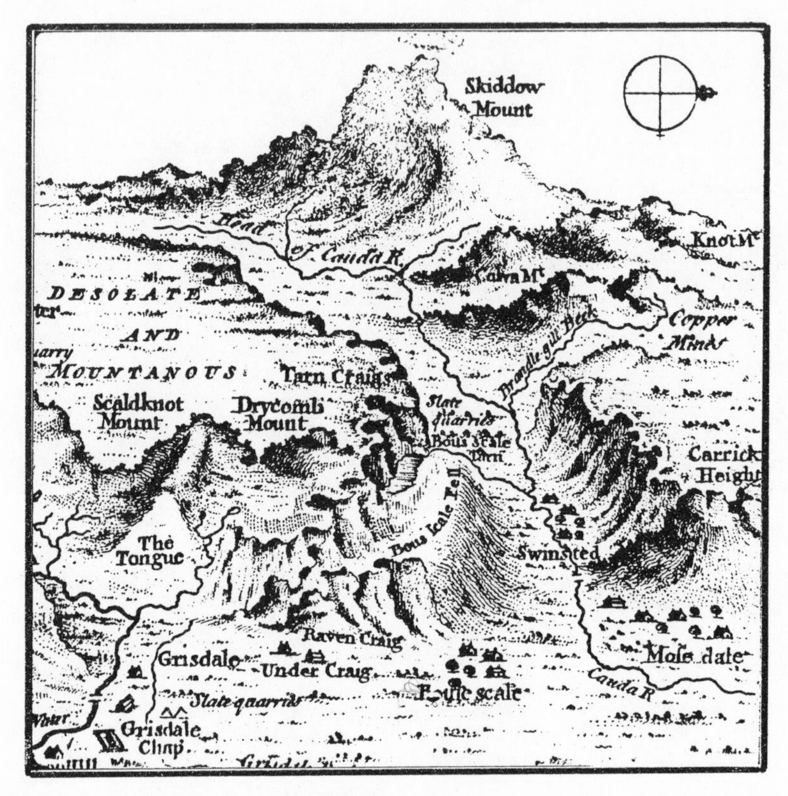

Figure 3 Map of 1747 showing Brandle-gill Beck.

beneath the place where it was found, continued their trials long enough to explain the consignment of bismuth sent to London and Antwerp around the end of January 1569?

The principal bismuth ore in these veins is again the sulphide bismuthinite,[43] but here elongated crystals up to 25 mm have been found in quartz,[44] associated with the sulpho-tellurides (principally josëite) and small amounts of native bismuth.[45] Fine specimens of the latter are said to have been found in Emerson's vein.[46] There are early open cuts on these veins, from 2 to 7 feet wide, in a highly fractured granitic bedrock which is weathered and in many places lichen-covered, and although the use of pre-powder techniques is not immediately evident they cannot be ruled out.

This attempt to pull together several apparently unrelated facts perhaps throws more light on the extent of the mining operations carried

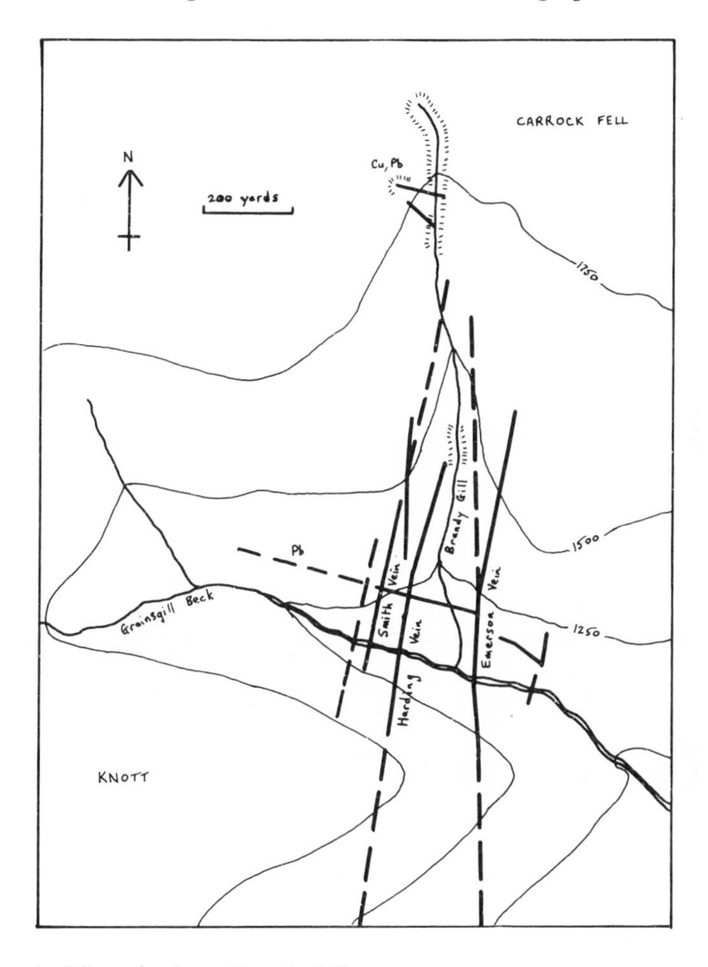

Figure 4 Mineral veins at Brandy Gill.

out by the Germans in Tudor times, though it is unlikely they would have found the rich silver deposits they appear to have been searching for. It is ironic that over three centuries later German miners William Boss and Frederick Boehm should have come to the same place to extract wolfram,[47] the very ore which in all probability baffled their predecessors.

EARLY USES OF BISMUTH

By the end of the sixteenth century an alloy containing bismuth and antimony was being used for casting type.[48] As well as having the advantage of a low melting-point and high fluidity, bismuth shares with antimony the unusual property of expansion on cooling, thus enabling the manufacture of alloys which undergo zero dimensional change on solidification,[49] a fact that may already have been recognized by the printers of the time. Moreover, small amounts of bismuth were occasionally added to the tin–lead alloy used in the manufacture of pewter to improve its hardness. Antimony was also employed for this purpose and the practice had been adopted in Europe before 1600, though apparently not in England until the late seventeenth century.[50] The first recorded use of bismuth compounds for therapeutic purposes, in particular for the treatment of digestive disorders, appears to have been in the seventeenth century, but its potential in this direction may well have been appreciated earlier, as well as its value as a cosmetic in the form of bismuth oxychloride, more commonly known as 'pearl white'.[51]

Notes and References

1. H. Hamilton, *The English Brass and Copper Industries to 1800*, 2nd edn (London 1967), 4–13.

2. W.R. Scott, *The Constitution and Finance of English, Scottish and Irish Joint Stock Companies to 1720, Volume 2* (Cambridge, 1911), 386.

3. W.G. Collingwood, *Elizabethan Keswick* (Cumberland and Westmorland Antiquarian and Archaeological Society Tract Series No. 8, Kendal, 1912; republished M. Moon, Whitehaven, 1987).

4. *New Encyclopaedia Britannica: Macropaedia.* 1991. 15: 974.

5. Alonzo Alvaro Barba, *El arte de los metales*, trans. R.E. Douglass and E.P. Mathewson (New York, 1923), 55.

6. Georgius Agricola, *De re metallica*, trans. H.C. and L.H. Hoover (New York, 1950), 3 (fn) and 433 (fn).

7. Agricola, *op. cit.* (6), xxvii (fn) and 610–12.

8. Agricola, *op. cit.* (6), 433–7.

9. E.S. Dana, *A Textbook of Mineralogy*, 17th edn (New York, 1893), 227, 284 and 412.

10. W.H. Dennis, *Metallurgy of the Non-ferrous Metals* (London, 1961), 521.

11. B. Young, *Glossary of the Minerals of the Lake District and Adjoining Areas* (British Geological Survey, 1987), 25–6.

12. Collingwood, *op. cit.* (3), *passim.*

13. W. Crookes and E. Rohrig, *A Practical Treatise on Metallurgy, Volume 2: Copper, Iron* (1869).

14. Kirk-Othmer, *Encyclopaedia of Chemical Technology, Volume 3*, 3rd edn (New York, 1978), 912.

15. J. Woodward, *An Attempt toward the Natural History of the Fossils of England* (London, 1729).

16. M.P. Cooper and C.J. Stanley, *Minerals of the English Lake District: Caldbeck Fells* (Natural History Museum, London, 1990), 65.

17. J. Percy, *Metallurgy: Fuel, Fireclays, Copper, Zinc, Brass, etc.* (1861).

18. Dana, *op. cit.* (9), 255. C.J. Stanley (1982), private communication.

19. Young, *op. cit.* (11), 17 and 98–9.

20. D. Grant, 'The Sixth Duke of Somerset, Thomas Robinson and the Newlands Mines', *Trans. Cumberland and Westmorland Antiquarian and Archaeological Society*, New Series, 1985, 85: 143–65, 153.

21. C.J. Stanley and D.J. Vaughan, 'Interpretative Studies of Copper Mineralization to the South of Keswick, England', *Trans. Institution of Mining and Metallurgy*, 1980, 89(B): 25–30.

22. C.J. Stanley and D.J. Vaughan, 'Copper, Lead, Zinc and Cobalt Mineralization in the English Lake District: Classification, Conditions of Formation and Genesis', *J. Geological Soc. London*, 1982, 139: 569–79.

23. K.C. Dunham, *Geology of the Northern Pennine Orefield, Volume 1: Tyne to Stainmore* (British Geological Survey, 1990), 69 and 72. K.C. Dunham and A.A. Wilson, *Geology of the Northern Pennine Orefield, Volume 2: Stainmore to Craven* (British Geological Survey, 1985), 88.

24. G. Hammersley, ed., *Daniel Hechstetter the Younger: Memorabilia and Letters 1600–1639*, Deutsche Handelsakten des Mittelalters und der Neuzeit, Band XVII (Stuttgart, 1988).

25. *Encyclopaedia Britannica*, 1947, 4: 670. Kirk-Othmer, *op. cit.* (14), 912.

26. M.B. Donald, *Elizabethan Copper: The History of the Company of Mines Royal 1568–1605* (London, 1955; republished M. Moon, Whitehaven, 1989), passim.

27. *Ibid.*, 251 and 293–4. G. Hammersley, 'Technique or Economy? The Rise and Decline of the Early English Copper Industry, *ca.* 1550–1660', *Business History*, 1973, 15: 1–31, 4–5).

28. Hamilton, *op. cit.* (1), 28 and 78.

29. Donald, *op. cit.* (26), 292–7.

30. Hammersley, *op. cit.* (27), 5 (fn). Donald, *op. cit.* (26), 294–5.

31. I. Abrahams, 'Joachim Gaunse: A Mining Incident in the Reign of Queen Elizabeth', *Trans. Jewish Historical Soc. Great Britain*, 1899–1901, 4: 83–101, 87.

32. Scott, *op. cit.* (2), 395.

33. Abrahams, *op. cit.* (31), 87.

34. *Ibid.*, 93–9.

35. Collingwood, *op. cit.* (3), *passim.*

36. Donald, *op. cit.* (26), 151.

37. Collingwood, *op. cit.* (3), 5. J.D.J. Wildridge (1993), private communication.

38. Hammersley, *op. cit.* (24), 153.

39. Anon., 'A Journey to Caudebec Fells, with a Map and Description of the Same', *Gentleman's Magazine*, 1747, 17: 522–5.

40. J. Adams, *Mines of the Lake District Fells* (Dalesman, 1988), 92–5.

41. Cooper and Stanley, *op. cit.* (16), 42–3. W.T. Shaw, *Mining in the Lake Counties* (Dalesman, 1972), 49.

42. W. Whellan, *The History and Topography of Cumberland and Westmoreland* (1860), 226.

43. E.H. Shackleton, *Lakeland Geology* (Dalesman, 1973), 71 and 126.

44. Cooper and Stanley, *op. cit.* (16), 84.

45. C.S. Hitchen, 'The Skiddaw Granite and Its Residual Products', *J. Geological Soc. London*, 1934, 90: 158– 200. R.J. Firman, 'Epigenetic mineralization', in F. Moseley, ed., *The Geology of the Lake District* (Yorkshire Geological Soc., 1978), 231. J. Hartley, 'A List of Minerals Associated with the Ore Deposits of the Caldbeck Fells, Cumberland', *Trans. Leeds Geological Assn*, 1984, 10(3): 22–39.

46. Shaw, *op. cit.* (41), 51.

47. D. Blundell, 'Wolfram', in *Beneath the Lakeland Fells* (Cumbria Amenity Trust Mining History Society, 1992), 109.

48. G.C. Boon, 'An Early Tudor Coiner's Mould and the Working of Borrowdale Graphite', *Trans. Cumberland and Westmorland Antiquarian and Archaeological Soc.*, New Series, 1976, 76: 97–132, 129.

49. *New Encyclopaedia Britannica: Macropaedia, op. cit.* (4). Kirk-Othmer, *op. cit.* (14), 912.

50. R.C. Hornsby, R. Weinstein and R.F. Hower, *Pewter: A Celebration of the Craft 1200–1700* (Museum of London, 1989–90), 15 and 47.

51. *New Encyclopaedia Britannica: Micropaedia,* 1991, 2: 242. Kirk-Othmer, *op. cit.* (14), 921.

Jean Errard (1554–1610) and His Book of Machines: *Le Premier Livre des instruments mathématiques méchaniques* of 1584

WALTER ENDREI

ABSTRACT

Although the books of machines of the sixteenth and seventeenth centuries are often discussed, Jean Errard's work, published in 1584 in Nancy, is rarely mentioned. Errard, a fortification engineer in the service of Lorraine who later went on to serve Henri IV, presents in his folio-sized collection of engravings, 40 in number, machine inventions both of his own and of others' devising. Especially interesting among his drawings are the first representations of spinning frames, the earliest drawing known of a horizontal rod-engine (*Feld-Gestänge*) and a mechanical saw powered by a windwheel.

It has become usual to assign the works of engineers to periods named after art-historical styles. So we find Bertrand Gille describing the technical drawings of the German and Italian schools, from Konrad Kyeser (1405) to the work of Leonardo da Vinci (d. 1519), as Renaissance technology.[1] The wonderful machine books of the period between 1550 and 1630 may be described with as much or as little justice as early Baroque works, although neither the intention nor the outcome of the representations appears to show artistic ambitions on the part of the draughtsman, except perhaps in the case of some of Ramelli's drawings of his inventions. Leaving aside any question of artistic ambition, if we look at the machine books without any such preconceptions, they are, just like the drawings of Kyeser or Leonardo, rational, expedient representations, the aim of which is to convey technical information, albeit information marked by the spirit and demands of the period. Artistic design assumes priority only exceptionally, as in the case of Besson's representation of a state coach. This we may place to the credit of his engraver, Andruet de Cerceau, rather than to Jacques Besson himself.

A special group of these technical books is remarkable for full-page engravings, mostly in small folio format, accompanied by relatively short explanations. In the case of other authors, near contemporaries of the machine books, such as Biringuccio, whose *Della pirotechnia* appeared in 1540, and Agricola, whose *De re metallica* appeared in 1556, the text carries considerably more of the burden of explanation. In Agricola, indeed, the work is admirably partitioned between word and image, something Biringuccio does not always manage. Sometimes in the work of Agricola the functioning of machines is made clearer by deliberate distortion on the part of the artist, or by showing machines in exploded form.

The species 'machine books', sometimes also called 'theatres of machines', the earliest work of which type bore the title *Theatrum machinarum*, has its origins in the Middle Ages. The technical manuscripts compiled by various architects and engineers from Villard de Honnecourt to Leonardo were not intended for publication. It was an exceptional case when, in the year 1474, the Prince of Urbino ordered that 72 of the superb machine drawings of Francesco di Giorgio Martini be cut in stone in relief, the panels to be placed on the outer walls of his palace. At about the same time there also appeared the first printed technical works to carry illustrations, those by Roberto Valturio on the art of war in 1472 and by Leon Battista Alberti on architecture in 1485. Thereafter the sequence is continuous. One may take as an outstanding work in this sequence the most famous technical book of the period, Agricola's *De re metallica,* which between 1556 and 1650 went through some 49 editions. In this work we find 292 woodcuts elaborately keyed to the text. In this way, text and illustration are thoroughly integrated and it would be difficult, for this and other reasons, to sustain the idea that one is subordinated to the other. The quality of the illustrations is self-evident. One of the artists was Zacharias Specklin, illustrator of the *Cosmographia* of Sebastian Münster. Similarly, some of the title pages of the machine books are themselves furnished with excellent graphics. For instance, Mathias Merian prepared the copper engraving for the title page of Jacobus de Strada's work. While the texts of the machine books try, with varying success, to explain the representations in the works, the purpose of engravings in other technical, but less image-reliant works – be they by Cardano, Salamon de Caus, or later the *Grande Encyclopédie* – are simply intended to amplify the text.

The genre of the machine books undoubtedly first came into fashion in France but achieved its splendid latest flowering in Germany where Jacob Leupold and his continuators, between 1724 and 1739, were to publish ten folio volumes containing in all some 500 engraved copper plates. Historians of technology include in this group of theatres of machines as prominent examples the following works:[2]

- Jacques Besson: *Theatrum instrumentorum et machinarum* (Orléans, 1569), followed by later editions, notably Paris, 1578, when it bore the title *Théâtre des instruments mathematiques et méchaniques*, 60 plates.
- Agostino Ramelli: *Le diverse et artificiose machine* (Paris 1588), 195 plates.
- Vittorio Zonca: *Novo teatro di machine et edificii per varie e sicure operatione* (Padua, 1607), 42 plates.
- Heinrich Zeising: *Theatrum machinarum*, I–II (Leipzig, 1607–12), 128 plates.
- Faustus Verantius: *Machinae novae* (Venice 1616/17?), 49 plates.
- Jacobus de Strada: *Künstliche Abriss, allerhand Wasser-Wind-Ross- und Handmühlen* . . . , I–II (Frankfurt aM, 1617–18), 100 plates.
- Giovanni Branca: *Le Machine* (Rome 1629), 77 plates.

These works are far from being equal in value. Feldhaus, for instance, curtly dismisses the work of Zeising as a comprehensive evisceration of the works of older authors. It has to be said, however, that originality is not the strong suit of these authors. In this regard Verantius deserves praise because he often mentions where he saw a technique for the first time. He was, by the way, a cleric. Among these works there is also great variation as to the credibility and practicability of the representations: for instance, the very last plate in Zonca's collection is of a perpetual-motion machine! Four of the above-mentioned books, those of Zonca, Zeising, Verantius and Strada, appeared posthumously, their authors consequently not being able to give finishing touches to their works.[3] In Strada's case the editor actually goes so far as to say that written explanations are omitted because they are considered unnecessary.

The machine books of Besson and Ramelli were works prepared in their old age. This was not the case with the work of Jean Errard (1554–1610), whose book *Le Premier Livre des instruments mathématiques méchaniques*, the subject of this paper, was published in Nancy in 1584 when he was only 30.

Neither Beck in his *Beiträge zur Geschichte des Maschinenbaues* (Berlin, 1899) nor A.G. Keller in his *A Theatre of Machines* (London, 1964) mentions Errard and his works. A limited facsimile edition, published in 1979, ensured that at last the work of 1584 would be available to a wider audience and not simply to those privileged few who could actually handle the volume themselves.[4] Apart from this Errard is further distinguished from the other authors by the fact that we are fairly well informed about his career. He was born in Bar-le-Duc, the capital of the princedom of Bar, at this time already under the rule of Lorraine. He studied in Italy the new art of fortification and subsequently entered the service of Charles III of Lorraine, to whom he dedicated the book of 1584. Later he took employment with the House of Bouillon, his employer there being Charlotte de La Marck, Princess of Bouillon.

By the late 1580s he was working as engineer to the King of France, perhaps the last of the Valois. What is certain is that he was in the employment of Henri IV (1589–1610) throughout his reign. It was the latter who nominated him as *premier ingenieur*, drew him into the Conseil Royal and in 1591 granted him the right to mint coins. Errard took part in the sieges of Chartres (1591) and Dreux (1592), and later fortified Sedan (1595), Montreuil-sur-mer (1597) and Amiens (1598). In 1599, after the peace treaty of Vervins, he was elevated to the nobility. In the peace that followed he superintended various fortification works. His other book, *La Fortification reduite en art et demonstré* (Paris, 1594), was translated into German and was the first work of this type written in French.[5] Errard died in the same year as Henri IV, 1610.[6]

Like the works of Besson and Ramelli, Errard's book can be esteemed as an achievement of French military engineering. In the wars of religion Besson served under the Valois and in the end suffered persecution as a Protestant. Ramelli, a Catholic, was captured during the siege of La Rochelle by the Huegenots. Errard's war experiences were subsequent to the publication of his book and occurred in the service of the Bourbons. Perhaps it is for this reason that there are few military subjects among the engravings in his book: only numbers 4, 5 and 7. Military matters he saved for his second book, the *Géometrie et pratique général* (Paris, 1594). This was a successful work and ran into two subsequent editions (1602 and 1619). It is worth mentioning that he published a translation of the first six books of Euclid in 1598 and wrote a pamphlet against J.J. Scaliger, who believed he had solved the problem of the squaring of the circle.[7]

Le Premier Livre des instruments was published in Nancy in early 1584 in the workshop of Jean Janson, the court printer of the House of Lorraine. The name of the engraver is not known. The dedication makes it clear that Errard had already been for some time in the service of Lorraine and that he had further drawings in the course of preparation: '*quime sont encor demeurez imparfaicts*'. There then follows a short passage in which, turning to the reader, he claims the machines to be his own creations, emerging '*premier de ma boutique*', while nevertheless admitting that it could easily happen that two people could hit upon the same idea.

Errard's concession raises the obvious question as to whether he knew Besson's work, which had by 1584 already appeared in two editions. Errard does mention, for instance, Cardano's *De subtilitate rerum*, first published in Frankfurt in 1550. Naturally, he also mentions the works of the Hellenistic engineers, such as Heron and Polybius, both of which were published in the course of the sixteenth century. What one would most like to know would be whether Errard had seen the 60 copper engravings of Besson's *Theatrum instrumentorum* . . . , published first in Orleans in 1569 or 1571 and then in Lyons in 1578 with a French title, with Latin explanatory texts by François Beroald. In the same year, 1578, two Italian editions appeared, in Lyons and Venice, and finally a reprint in 1582.[8] It is difficult to believe that the library of the Prince of Lorraine in Nancy did not contain at least one of these five editions. In any case, other indications allow one to conclude that Errard knew the work of his predecessor. First of all, there is the fact that Errard's title on the title page of his book was given in Latin but was also supplied with a French translation, just as was the case with Besson's book. Curiously enough, in the case of some of his engravings, Errard was at pains to point out who was responsible for the invention (see plate numbers 19, 25 and 29 and explanatory texts), yet never mentions Besson. At the same time the knowledge of the model created by Besson can be supposed in several instances, although it is possible, of course, that both may have drawn on some earlier source, as, for example, the turning lathe (plate number 3 in Errard's work) and machines numbers 7, 8 and 9 in Besson's collection. It is obvious that Errard borrowed from others too, notably the idea of the Archimedean screw and water-wheel of plate 19, which has an obvious source in Cardano. Finally, one should mention here that the idea of a vertical-axle windmill (plate 19) was presented in a similar manner by Gualtherius Rivius in 1547,[9] which itself can be traced back to a drawing of Mariano Taccola of *c.* 1438 (see Figure 1).

Another group of plates deals with technical problems that Errard should have known had been solved in practice long before. It is difficult to understand how he could ever have claimed these as his own inventions. In this category belong most of his devices for lifting and transporting materials, such as numbers 1, 2, 5, 6 and 9, the horizontal power-transmitting device (21), the pound lock (24) and the mills (13 and 14). One can, of course, observe that in these examples some small modifications have been introduced, although usually to the detriment of the machine concerned. So, for instance, the hauling device shown in number 5 deploys a chain driven by a lantern gear for the purposes of moving a cannon. For this purpose a sprocket chain would have been necessary.

Something further should be said here regarding Errard's engraving (21) of a rod-engine. Here a pump at some distance from the water-wheel, the power source setting it to work, receives its motion via a reciprocating rod line suspended from a series of supporting legs. Such devices were almost certainly coming into use in the great centres of European mining during the 1570s. The longer the transmission line, the greater is the loss of energy through friction, although the suspended rod idea is plainly an attempt to minimize such losses. The famous water-lifting devices of Toledo built by Juanelo Turriano were scarcely able to deliver more than what one Spanish critic long ago called an '*exigua dotacion*', although there is no reason to suppose that the mining machines themselves were anything other than successful. A critical assessment of plate 21 and its place in the history of machine development up to the end of the sixteenth century throws some further light on the nature of Errard's contribution in this instance. In short, his rendering of the machine would ensure its prompt breakdown, since such machines could not function in the double-acting fashion that Errard is proposing here. At all events, this machine by means of which '*on puisse tirer facillement l'eau d'une source assez estoignée*' could not have been an invention of Errard's.[10]

What is perhaps much more interesting to investigate is those ideas which, even if they are not original ideas and inventions of Errard himself, are nevertheless the earliest representations we have of hitherto unknown techniques. In the first place, we see what appear to be the earliest experiments designed to mechanize spinning or to devise a spinning device with several spindles, as in plate 25 (See Figure 2). Here, Errard indicates correctly that the machine is not of his own devising, for the honour is due to 'maistre Charles Desrué' (Charles Deruet), who was

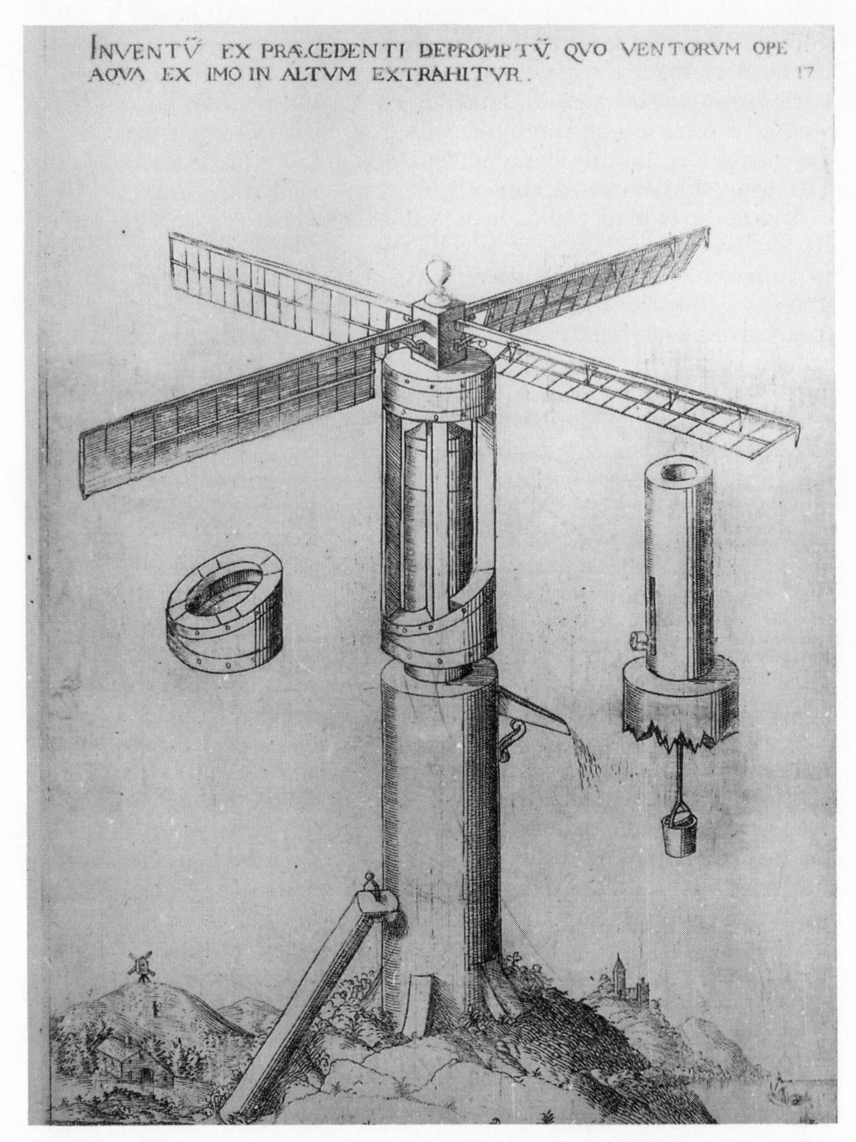

Figure 1 Errard's Latin caption runs, 'An invention developed from those previous by which the wind lifts a great quantity of water from a low place to a high one.' Note the self-luffing sails of the windmill.

experienced in such subtle inventions. Deruet seems also to have been responsible for number 29. The machine in question seems to be derived from the silk-throwing mills of Lucca, even if it is not, unfortunately, practicable. The yarn runs, just as in the case of the silk-throwing mill, from the spindle to the winding-up device, to the reeler. Likewise, plate 26 represents a further development made by Deruet, with possible additions by Errard, who may have given it a horizontal layout.

Both proposals are based on the fact that the carded wool is placed in small receptacles, or cops, through the openings of which the wool is pulled out uniformly (just as a hand spinner would do at a spinning wheel), is given a twist and is then led to the reeler or the spindle, and is there laid down. However, the spinner carries not two tasks but three: she also has to ensure the drawing out of the roving, that is, the fore-yarn (now called 'rover' by technologists), preparatory to spinning proper. This procedure was mechanized only in the eighteenth century by means of pairs of drafting rollers, the second pair revolving more quickly than the first pair and thereby drawing out and thinning down the roving.[11]

The experiment to solve the problem of driving a mechanical saw by a wind-wheel also probably emerges for the first time in Errard's book (30). This invention is ascribed to the Dutchman Cornelius van Uitgeoot in 1592. The first documented reference to such a saw-mill comes only in the year 1633.

W. von Stromer has described the metal-cutting device shown in plate 29 (Figure 3) as a novel idea, and pointed out that in terms of inventive value it is of exceptional worth. He notes that almost certainly this intimation of material forming by means of steel rollers reveals a totally new concept. The water-powered machine drives two pairs of steel discs, one offset from the other, with the effect that their edges, acting against each other in opposite directions, cut like scissors. Errard's picture shows us that this hitherto unknown method of slitting iron from a thick plate is scarcely different from that used in the steel mills of today.[12]

While the invention above is ascribed by Errard to Desrué/Deruet, the cylinder printing machine figured in plate 20 might come in reality from Errard's own *boutique* (see Figure 4). Employing a pair of cylinders in printing books in place of the screw press was often the subject of proposals after 1600.[13] It is not altogether clear how the frame rolls on the printing table are to be moved in Errard's representation. Such deficiencies can, however, be solved without difficulty, and the picture is surely the earliest representation of such a radical departure from traditional printing methods.

We turn next to the sailing ship equipped with paddle-wheel shown in engraving number 23 (see Figure 5). The caption above the representation of the pumps shown in the left-hand corner of the engraving makes it clear that the paddle-wheel, turned by the motion of the ship, drives these pumps, whose purpose is to remove water from the bilges of the ship. Suction-lift pumps, or displacement pumps, driven by wind-wheels as shown in engravings numbers 17 and 18, might be the oldest representations of this kind.

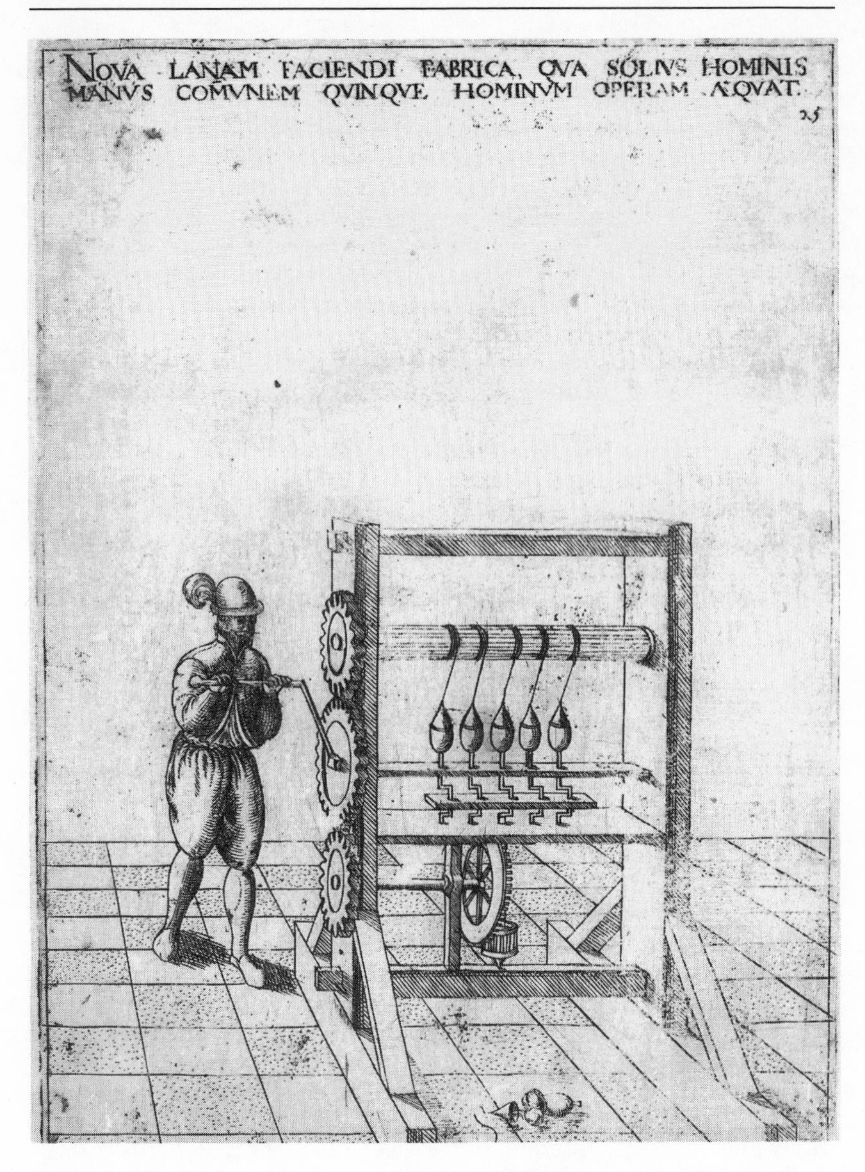

Figure 2 The Latin caption runs, 'A new device for spinning wool by which one man alone equals the work normally done by five.' Errard remarks in his French description of this machine that credit for it belonged to Charles Desrué, a man skilled in subtle inventions. In fact, the Deruet family were clock-makers and therefore well versed in mechanics. It may be noted here that Errard credits Deruet with also being the first to demonstrate iron-slitting, as shown in Figure 3.

Figure 3 The Latin caption runs, 'A new machine for slitting iron hitherto unknown.'

Quite original is Errard's hand-driven tilt hammer (27). What is novel here is his use of a flywheel with built-in flyballs to facilitate the work. He does not conceal the fact that he got the idea '*de ceulx qui pensent par telz contrepois avoir trouvé . . . le mouvement continuel*'. Our author might have picked up the idea of fitting out the chimney of a forge with a boiler in order to supply the smith's bellows with a '*continuel vent*' in place of the discontinuous blast yielded by ordinary cuneate bellows (32) from the then well-known sufflators or *Püstriche.* Equally, the idea has some affiliations with the trombe or catalan forge blower.[14]

Errard also designed bridges, as one may see in engravings 32, 33 and 34. The use of stone elements of the design shown might have come from beam structures built with scarfed joints, such as that figured by Verantius as the Pons Ligneus.

Finally, we should mention Errard's four pedometers, shown in plates 37–40. Leonardo had previously designed pedometers for use with horses and pedestrians. In both cases the crucial ratchet wheel is either mentioned or drawn. Long before this Heron (second half of the first century AD) had described hodometers on carriages, and devices like this were proposed several times in the course of the sixteenth century. A properly functioning log, i.e. a hodometer for ships, was, however, very badly required. Although a ship's log with knotted cord had been devised by Humphry Cole in 1577, neither Errard nor Besson before him seemed

Figure 4 The Latin caption runs, 'A new kind of press for printing what you will more cheaply and easily than by other means.'

to have known of it. Besson, indeed, in his fifty-seventh figure, had proposed a measuring tool driven by a paddle-wheel which would obviously function only when it was immersed in water, something not evident from the representation. The device designed by Errard, by contrast, deploys a horizontal wheel with vertical axis. One may ask whether, in this case, Errard had done any more than modify Besson's idea so as to give it some semblance of novelty.

Errard's work seems not to have made the least impression on his contemporaries. The work of Joseph Ballot, also rare, entitled *Modelles artifices de feu et divers instruments de guerre*, published in Chaumont in 1598, is thought by France-Lanord to have been influenced by Errard's work. This perhaps may also be true of the later authors, Jean Appier and Francois Thybourel.[15] At any rate, Errard's book certainly deserves to be more carefully studied as a source for the technology of the sixteenth century.

Revised translation from the German by Graham Hollister-Short

Notes and References

1. B. Gille, *Les Ingenieurs de la Renaissance* (Paris, 1964).
2. The list may be extended to the time of Jacob Leupold (d. 1727) but discussion here will be limited to some only of Errard's contemporaries.

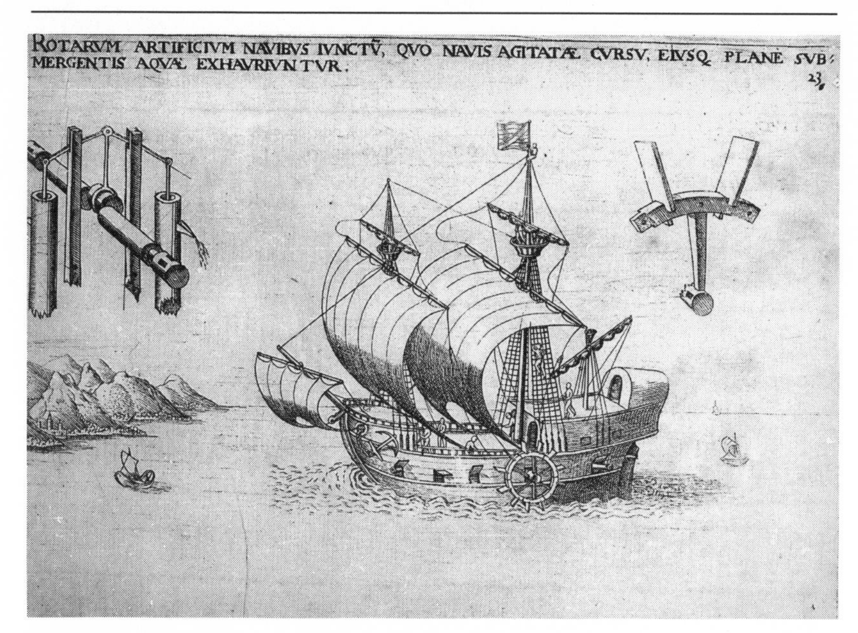

Figure 5 The Latin caption runs, 'Bilge waters are plainly pumped out by means of ingenious wheels fitted to ships, which wheels are moved as the ship makes passage.'

3. With Besson the case is somewhat different. A Protestant, he fled France to escape persecution and died either in London (1573) or Geneva (1576). Subsequently his work fell into the hands of a François Beroald who produced the edition of 1578 (falsely claiming it to be the first edition). The plates have Latin captions, like Errard's later, and are described in French too (again like Errard's). For further information on Besson see two papers by A.G. Keller: 'The Missing Years of Jacques Besson', *Technology and Culture*, 1973, 14; and 'A Manuscript Version of Jacques Besson's *Book of Machines* . . . ', in *On Pre-modern Technology and Science. Studies in Honor of Lynn White Jr*, ed. Hall and West (Malibu, 1976). I have to thank Graham Hollister-Short for bringing these references to my attention.

4. Only three copies of Errard's work of 1584 are known to exist, so when Albert France-Lanord speaks of only a privileged few ever having had it in their hands, he is correct. France-Lanord's facsimile reproduction was published in Nancy in 1979, using the Nancy copy.

5. This work was re-published in revised and enlarged form in 1620 by A. Errard, Jean Errard's nephew.

6. Unlike other authors of machine books considered here, a great deal of information is available relating to Errard's life. For further details see M. Lallemand and A. Boinette, *Jean Errard, les oeuvres, sa vie, sa fortification* . . . (Paris, 1884). I have not been able to consult this book.

7. A. France-Lanord, *op. cit.*, introduction, x. One might notice here how curious it is that neither Errard's book nor Besson's contains very many pictures relating to military technology. Ramelli's, by contrast, contains no fewer than 25.

8. F.M. Feldhaus, *Die Technik. Ein Lexikon* . . . (Berlin, 1914, rpt 1970), column 82, sub Besson. France-Lanord gives 1569 for the first edition of Besson's work, as do several entries in the *National Union Catalogue*.

9. T. Beck, *Beiträge zur Geschichte des Maschinenbaues* (Berlin, 1899), 184. See also G.H. Rivius (W.H. Ryff), *Der furnehmbsten nothwendigsten der gantzen Architektur gehörigen mathematischen und mechanischen Kunst eygentlicher Bericht*... (Nuremberg, 1547), III, 41.

10. On the problem of horizontal transmission of power by means of reciprocating horizontal rod lines (*Feld-Gestänge*) see G. Hollister-Short, 'The First Half Century of the Rod Engine (*c.* 1540–*c.*1600)', *Peak District Mines Historical Society Bulletin*, 1994, 12: 83–90. See also W. von Stromer, 'Wassernot und Wasserkünste in Bergbau des Mittelalters und der frühen Neuzeit', in *Montanwirtschaft Mitteleuropas vom 12 bis 17 Jahrhundert*, ed. W. Kroker and E. Westermann (Bochum, 1984), Beiheft 2, 63. Stromer remarks here that the first exact depiction of a horizontal rod engine appears in Nuremberg in 1597–8. The drawing of the device which was used for de-watering caissons during bridge construction is to be seen in the archives of the *Schloss Grünsberg*.

11. As in Arkwright's patent of 1779 for roller drafting. A. France-Lanord is in error in seeing something similar in Zonca's work. Zonca's textile machines are, however, for throwing silk.

12. W. von Stromer, 'Apparate und Maschinen von Metallgewerben in Mittelalter und Frühneuzeit', in *Handwerk und Sachkultur*, ed. H. Kühlen (Vienna, 1988), 147, Fig. 30.

13. See, for example, Zonca's *Novo teatro* of 1607, 76, apropos copperplate engraving.

14. F.M. Feldhaus, *op. cit.*, col. 844, sub *Püstriche*.

15. A. France-Lanord, *op. cit.*, xiv.

From the Imperial-Royal Collection of Manufactured Products to the Museum of Technology and Industry in Vienna

HELLMUT JANETSCHEK

ABSTRACT

The beginning of the collections preserved in the Museum of Technology and Industry in Vienna dates back to the year 1807 when, during the reign of Francis I, the so-called Nationalfabriksproduktenkabinett was founded, a permanent display of samples of products from all parts of the Habsburg monarchy. In 1814 this collection was transferred to the Kaiserlich-Königlich Polytechnisches Institut (today's Technical University of Vienna), which claimed to be not only an education institution but also a museum of technology. The Fabriksproduktenkabinett was continuously enlarged and supplemented by other collections, in particular the Emperor Ferdinand's private collection, and ended by being called the Technologisches Kabinett in 1842. In 1912 the greater part of the Technologisches Kabinett was – simultaneously with other collections – transferred to the Museum of Technology and Industry. The founder of the museum (opened in 1918), as well as of some forerunner institutions, was Wilhelm Exner (1840–1931), who was educated at the Polytechnisches Institut. It was he who also determined the main principles of organization of the museum, which remained in force until its closure in 1992.

INTRODUCTION

Museums of technology (not science centres) with their accumulation of artefacts are nowadays more and more acknowledged as resources for the history of technology, and feel an obligation towards this discipline, which they see as an essential part of the history of human culture. Corresponding to the tradition of the collections that went into its formation, the Museum of Technology and Industry in Vienna can be ranked as one of the oldest in the world, and, in terms of the extent and quality of its collections, as one of the most important too. Unfortunately, at the time of writing, these collections will not be open again to the public for many years. The museum was closed on 1 September 1992, and its buildings are scheduled to undergo reconstruction from the summer of 1994 to the spring of 1997. The director at the time of closure, Peter Rebernik, and the authorities in charge decided to close all the collections in their entirety. Of course this closure could not take place without serious interruption to scholarly research. Since the museum will only reopen at the earliest in spring 1998 (and then of course only in stages) on an entirely new plan, the author asks the reader to understand that this report relates necessarily more to the museum's past than to its future. The normal course of being able to check objects against records and literature is not possible at the moment, although the archives and library are still accessible to the public in their present accommodation.

THE BEGINNINGS OF THE COLLECTIONS

The basic stock forming the collections of the Museum of Technology and Industry was established in 1807 with the foundation of the so-called Fabriksproduktenkabinett,[1] which was created by order of Francis I. The Imperial order stated that

> a special cabinet is to be set up such as will enable everyone to obtain a general view of what is manufactured in the Imperial lands, thereby promoting sales and stimulating industry. From the total estate of home-made products and manufactured goods those which are excellent, good, suitable, beautiful, tasteful and instructive, in not too small dimension, with mere varieties excluded, are to be gathered in a specially chosen building to be well lit, related objects to be harmoniously arranged . . . this magnificent museum is to be enriched systematically with selected tools, instruments and books. Public exhibitions, the awarding of prizes and the organizing of prize competitions are to be staged too. In a word, a public institution is to be created which, raised to the desired level, would be the pride of the nation, a stage on which the country's artistic industry could be displayed, a model school for artists and craftsmen, an object of admiration for foreigners and of reproach for unpatriotic citizens, the blind admirers of foreign work.

Without conceding free trade, the cameralistic[2] economic policy of the Habsburgs thus took up principles which were at that time also practised in enlightened and revolutionary France.

During the time of the Empress Maria Theresa lectures had already been given at the University of Vienna, which, illustrated by the collection of models[3] assembled by the Professor of Mechanics, Walcher, popularized modern methods of production. These lectures were open to tradesmen as well as craftsmen. In the reign of Joseph II the sales depots of factories (that is, enterprises not belonging to any guild) made samples available for distribution to tradesmen. A notable event took place in 1791, when the first trade exhibition to be staged on Habsburg territory was held in Prague on the occasion of Leopold II's coronation as King of Bohemia.[4]

Aloys von Widmannsstatten (1754–1849),[5] a former manager of Austria's first power-driven spinning mill, a man of wide technical knowledge,was designated director of the Nationalfabriksproduktenkabinett. At least the first step had been taken: the Kabinett was housed in Vienna. In 1814 the Emperor ordered that the Physikalisches Kabinett, founded by Franz Stephan, and the Fabriksproduktenkabinett should both be housed on the site of the Polytechnic Institute (today's Technische Universität Wien).[6] This was opened in the following year. It was to be the second oldest institution of its kind in the German-speaking lands, and became a model for the whole of Germany (Figure 1). The director of the new institute was Johann Joseph Prechtl (1778–1854).[7] In the constitution of the Institute,[8] largely his work, it was definitively laid down that it was to serve as:

1. An educational establishment.
2. A museum of technology or conservatory for arts and crafts.
3. An association for promoting national industry or society for the encouragement of arts and crafts.

Figure 1 K.k. Polytechnisches Institut in Vienna about 1920.

Together with the collections of the preparatory Realschule (with collections relating to mineralogy, zoology and geography) and of the commercial section (with collections of merchandise), the collections summarized in 2 above contained the core collections of the technical section. These were: (a) a collection of chemical preparations and products; (b) mathematical instruments; (c) instruments for the study of natural (i.e. physical) philosophy; (d) models (Figure 2); (e) a mathematical and mechanical workshop; and (f) the Fabriksproduktenkabinett (Figures 3 and 4). In line with the institute's constitution the collections of models were placed in the care of the Professor of Practical Mechanics, the models coming from the collections of the Imperial family, from purchases made from domestic and foreign mechanics[9] and from the institute's own workshop.[10]

In 1816 the Fabriksproduktenkabinett was assigned to the new Professor of Production Science (Mechanische Technologie), Georg Altmütter (1787–1858),[11] who in the German-speaking lands is considered the founder of this discipline. Altmütter also established the first systematic collection of tools.[12] This was integrated into the Fabriksproduktenkabinett and supplemented with models. The Kabinett

Figure 2 Model of a Trevithick steam engine made by Johann Dietrich Langenreiter in 1804. The original had been built in 1803 by David Watson in London and sent to Eisenstadt, the Hungarian residence of Prince Esterházy. His mechanician, Langenreiter, installed this engine in the park of Esterházy Castle to supply a pool and to heat a bath. The model was transferred in 1816 from the Emperor's collection to the Polytechnic Institute.

comprised about 6000 items in 1816 and about 30,000 by 1829. The constitution of the institute stipulated that only products of the best possible quality were to be collected. These were arranged chronologically according to the different branches of industry. These not only

Figure 3 Drinking cup with portrait of Emperor Franz cut by Joseph Piesche, Pest (Hungary) 1824. Although the centres of Austrian glass manufacturing were located in Bohemia and the woodlands, subsequent treatment was often carried out in the towns.

Figure 4 Iron 'cast-tree' for costume jewellery by Joseph Glanz (1795–1866), 1831–49, Vienna. Glanz was born in Lemberg and educated by Christian Daniel Rauch at the Royal Academy in Berlin, where he also worked at the Royal ironworks. In 1831 he established a workshop in Vienna producing manufactured cast products of high quality.

served as aids in the teaching of technology but also provided material for the series of annual exhibitions. These latter finally led to the Vienna trade exhibitions of 1835, 1839 and 1845, organized to a great extent by the professors of the K.k. Polytechnic Institute, who from 1851 onwards were in much demand as *rapporteurs* at the world exhibitions. During the first half of the nineteenth century they also had to adjudicate on applications for Austrian patents, or for privileges, as they were called. The original records of these patents are kept at the Technical University of Vienna. Only some few of the corresponding patent models survive at the Museum of Technology and Industry. For comparison, not only domestic products but also foreign manufactures were taken into the Kabinett.

In 1842 the collection of products formed by Emperor Ferdinand and called the Technisches Kabinett, founded in 1819, and gathered by the Inspector of Factories, Stephan Ritter von Kees (1774–1840),[14] was transferred to the Polytechnic Institute and united with the Fabriksproduktenkabinett and the collection of models. The united collection was henceforth called the Technologisches Kabinett. Later, parts of this latter Kabinett were transferred to the newly established Museum of Art and Industry (now the Museum of Applied Art), where they are still preserved. In 1892 another part of the collection was transferred to the Museum of History of Austrian Labour, which was itself associated with the Technologisches Gewerbemuseum. The stock of both these last-mentioned museums was transferred in 1912 to the Technisches Museum für Industrie und Gewerbe simultaneously with the collections of the Technologisches Kabinett, up to then still at the Technical University. Somewhat less numerous, but also begun in 1807, is the collection of models (Figure 5) of the Society for Agriculture (K.k. Landwirtschaftsgesellschaft).[15] This well-documented collection, because of its importance for the history of agriculture, was also transferred to the Museum of Technology and Industry from the Hochschule für Bodenkulture (University of Forestry and Agriculture) in 1912.

OTHER FORERUNNERS OF THE MUSEUM OF TECHNOLOGY AND INDUSTRY

Because of changes in economic and social life brought about essentially by technology, the attitude towards trade exhibitions of technology changed, as did the attitude towards museums. Academic schools of technology, being primarily concerned with contemporary education, could no longer satisfy the demands posed by the need to conserve and present historical material; nor could such demands be satisfied by trade or industrial exhibition organizers. In the person of Wilhelm Franz Exner (1840–1931)[16] a man arose in Austria who uniquely not only became an initiator and renewer of technical education and of technical exhibitions but also developed the idea of museums of technology. This outstanding organizer was trained as an engineer at the Polytechnic Institute and qualified as a teacher and expert in industrial production, later becoming professor at the University of Forestry and Agriculture.

When only 19 years old he organized the hundredth anniversary celebrations of Schiller's birth in Vienna, and at 21 wrote the first history of the K. k. Polytechnisches Institut. He also wrote the first biography of the famous technological historian Johann Beckmann.

Exner, in addition to teaching and lecturing, was especially interested in exhibitions, and in 1816 published a booklet on the subject. He soon acquired an international reputation as *rapporteur* at various exhibitions in the German world, and, internationally, at the second Paris World Exhibition of 1867.[17] At the Vienna World Exhibition he not only organized the Austrian section but also put on an additional exhibition, a historic representation of Austrian industry and technology. Exner was well qualified for this work. In addition to the catalogue he prepared for the Vienna exhibition of 1873, he also wrote a book on the subject called *Beiträge zur Geschichte der Gewerbe und Erfindungen* (Essays on the History of Manufactures and Inventions). The additional exhibition referred to above was meant to buttress the Athäneum, conceived as both a school of continuing education for tradesmen and a museum. However, the Athäneum ceased to exist shortly afterwards because of lack of support and the departure of the organizer of the Vienna exhibition, Schwarz-Senborn, as ambassador to the USA. The oriental productions acquired at the Vienna exhibition fared better. They were transferred to the

Figure 5 Model of James Smith's mowing machine made in 1816 by Abbé Harder (b. Brühl/Gunsburg 1769, d. Vienna 1857) corresponding to the drawing and description in the *Encyclopedia Britannica.* Experiments were carried out in Austria in 1817 but with no practical consequences. Abbé Harder was model maker of the K.k. Landwirtschaftsgesellschaft from 1813 until 1855. During this time he produced more than 1,000 models.

Oriental Museum and later to the Museum of Trade, the important pieces thereafter going to the Museum of Art and Industry.

All this was still a long way from a museum for the history of technology. Exner, who tried to stimulate industry and trade primarily by education, never lost sight of the idea of such a museum. He tried to promote education by establishing vocational schools and creating technical colleges. With help from the Lower Austrian Industrial Union (the Niederösterreichische Gewerbeverein) and the benevolent support of the Palace, he was able to establish in 1879 the Technologisches Gewerbemuseum, the first technical college in Austria. As the name suggests, the school was from its beginnings connected with the museum. The college was housed at first in the Niederösterreichische Gewerbeverein and then moved to a disused locomotive plant. In line with Exner's interests, the museum emphasized wood technology and experimental research. Again, with the help of the Niederösterreichische Gewerbeverein, Exner in 1892 founded the Museum for the History of Austrian Labour (Museum der Geschichte der österreichischen Arbeit). This museum was closely connected with the Technologisches Gewerbemuseum and was housed in the same building as the latter.

Almost at the same time that the Museum for the History of Austrian Labour came into existence three other historical museums of technology were established. These were the K.k. Eisenbahnmuseum (Railway Museum) (1883), the K.k. Post- und Telegraphenmuseum (Post and Telegraph Museum) (1889) and the Gewerbehygienisches Museum (Museum of Industrial Hygiene) (1890). These three museums were state-run, in contrast to the Technologisches Gewerbemuseum and the Museum der Geschichte der österreichischen Arbeit, which were privately financed. Exner tried in vain to unify all five museums.

THE IDEA OF A TECHNISCHES MUSEUM FÜR INDUSTRIE UND GEWERBE AND ITS FINAL REALIZATION

It was only with the foundation of the Museum of Technology and Industry that all the five museums mentioned above were finally unified. The occasion for the unification was to be the sixtieth anniversary, in 1908, of the reign of Emperor Franz Josef. Various proposals were put forward but finally it was decided to proceed with Exner's long-existing plans for a museum of technology and industry, thereby creating an institution of permanent value.[18] To this end a preparatory committee was set up in 1907 made up of leading industrial figures, not least among them Exner himself, to carry the project forward. Exner played a leading role and determined to a great extent the structure of the different subcommittees and planning groups. This was work for which Exner was eminently well qualified since he had proved his organizing ability at several world exhibitions. On his own initiative and earlier than anyone else, he had published a proposal for just such a museum, including ideas for its organization, funding and ethos.[19]

The basic programme for a museum of technology geared to the needs of the new century had long been in Exner's mind. On his foreign travels he had looked closely into the organization of the Conservatoire des Arts et Métiers – an idea of the French Enlightenment – and Prince Albert's museum in Kensington. Though he was an admirer of both these institutions, he judged them to be too narrowly and nationalistically focused. Nor, although he had been involved in the foundation of the Deutsches Museum in Munich, did he think in terms of a museum of master works. For Austrian needs he thought that the purpose of a technological museum should be not only to explain technology in a didactic manner but also to place that technology in its social and cultural setting. Further, in order to engage the interest of an educated public who were less technically informed than the English or the French, but who were alive to the values of classical antiquity,[20] he was anxious above all to create a museum that would demonstrate the aesthetic qualities of engineering.

In his 'Main Principles for the Establishment of the Museum of Technology and Industry'[21] Exner maintained that the technical development of industry and crafts should be graphically demonstrated. Not only should the main aspects of industrial development be presented to the layman in a comprehensible manner, but examples of the latest developments in technique relating to particular industrial applications should also be acquired. The planning stage of building construction[22] should take into account the systematic future expansion of the collections. Materials should be primarily of Austrian origin, those of foreign provenance to be acquired only if they had been of importance for the industrial development of Austria. Strongly influenced by Darwin's theory of evolution, Exner thought it necessary to show not simply single objects but objects in their developmental sequences, if this were possible. In this way entire manufacturing processes and their historical development could be displayed. If machines or engines which were important links in these developmental lines were unavailable, then the museum should acquire reproductions. The library itself should correspond to the evolutionary historical character of the museum. To this end a collection of technical drawings should be undertaken. To popularize the museum, lectures on aspects of technology, industry and trade were to be scheduled. Last, but not least, Exner's conception of the Museum of Technology provided for the classification of all the collections into 13 main groups with their subdivisions. This idea was in general taken up by the organizing committee. Generally speaking, the guiding principle of the museum until its closure on 1 September 1992 continued to be based on Exner's conception, briefly enunciated above.

For the museum to become a reality, and for its construction to begin, the Committee of Finances under its President, Hugo von Noot (an industrialist), had to acquire funds. By 20 August 1908 it had collected 1,102,724 Crowns towards the estimated cost of some six million crowns. The government had promised to provide 1.5 million Crowns, while the municipality of Vienna gave one million Crowns in cash as well as donating the building plot (opposite Schonbrunn Palace) of about the same value.These financial provisions permitted the great work to begin.[23]

Exner's conception included a projected design for the museum building by Emil von Förster.[24] Time was short, and Förster's scheme would certainly have been executed had he not died in 1909; this, despite pressure from the Österreichische Ingenieur und Architektenverein (Union of Austrian Engineers and Architects) for an official competition.[25] His death obliged the construction committee to hold a competition. A precondition was that the total cost of the first phase of building should not exceed 3.6 million Crowns. Twenty-four Viennese architects, some of them with international reputations, such as Max Freiherr von Ferstel, Adolf Loos and Otto Wagner, took part in the competition. Most of them ignored the precondition. In the end the project was awarded to the architect Hans Schneider because his proposal was closest to Förster's scheme, which had been taken to an advanced stage of development in collaboration with the construction committee before Förster's death. The solemn laying of the foundation stone by Emperor Franz Josef took place on 20 June 1909 (Figure 6). In accordance with the wishes of the construction committee the plan (Figure 7), of which one-third only would finally be completed, was altered once more. In December 1909 Schneider was formally put in charge of the building works, although these in fact began only at the

Figure 6 The laying of the foundation stone on 20 June 1909. The speaker of the working committee delivers an address to the Emperor.

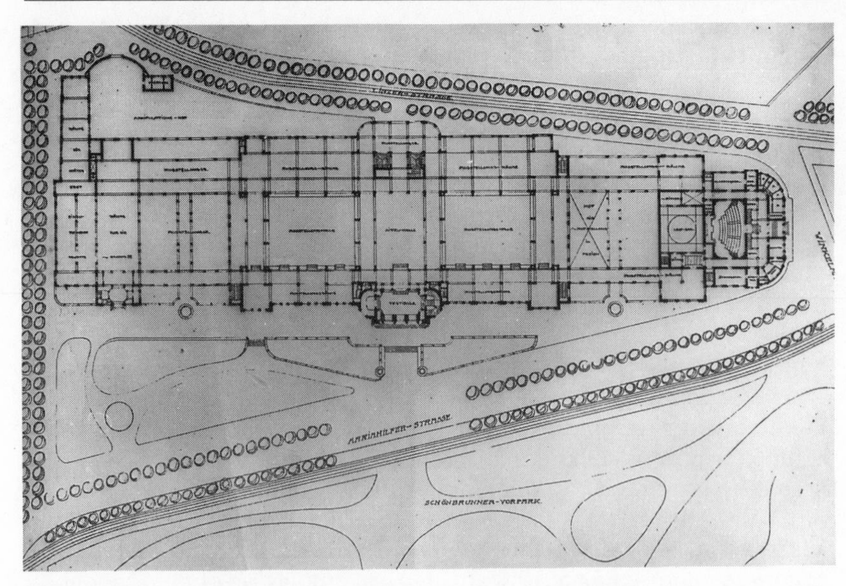

Figure 7 Finally altered plan of the museum building by Hans Schneider. Only the central part without the side wings was executed.

beginning of June 1910. The shell of the museum, one of the first reinforced concrete constructions in Austria, was completed before the winter of 1912–13. On 15 June 1913 the museum was complete enough

Figure 8 The museum building just before opening.

for the general meeting of the museum association (Museumsverein) to take place in the new building (Figure 8).

THE EQUIPMENT OF THE MUSEUM

When Exner's conception for the collections was adopted (that is, to acquire objects and organize them into 13 groups), the organizing committee recruited some 450 special consultants (later to be increased to 875) to oversee the work.[26] Although Exner was the leading spirit, there was need for a person whose exclusive task would be to organize the co-ordination of planning, ordering equipment and matters pertaining to display. The organizing committee found such a man in Ludwig Erhard (1863–1940),[27] who served as director of the museum from 1913 until 1930. Erhard, an engineer interested in the cultural aspects of his profession, before being called to Vienna by Exner to join the Gewerbe-förderungsdienst (Office for the Encouragement of Trade) had studied at the Technische Hochschule München (Technical University of Munich). He had afterwards worked at the Museum of Trade in Nuremberg and then in Bavaria. On 1 July 1909, shortly after the ceremonial laying of the foundation stone, Erhard was appointed technical counsellor by the working committee (Arbeitsausschuss).

Erhard now turned to the task of acquiring and sorting materials for the collections with vigour. An important step was taken in 1912 when several collections in the possession of official institutions were transferred to the museum. These were the stocks of the Gewerbehygienisches Museum, the Technologisches Kabinett (19,000 objects in the keeping of the Technical University), the Museum der Geschichte der österreichischen Arbeit, the Department of Saltworks, collections of weights and measures from the various departments concerned and the collection of agricultural models from the University of Forestry and Agriculture.[28] It was not until 1914 that the collections of the Railway Museum and the Museum of Posts were relocated to the Technical Museum, although they retained their own administrative structure until August 1980. The numerous objects donated by industry, meanwhile, were stored in the Rotunde, the building that had housed the Vienna World Exhibition of 1873. This was burnt down in 1937.[29]

Exner's conception of 13 industrial groupings was increased to 17 in 1913, and in the following year it was possible to begin deploying materials in the museum along these lines. The 17 groups were: (1) forestry and agriculture; (2) mining industry and metallurgical engineering; (3) iron and metals; (4) mechanical engineering; (5) electrical engineering; (6) vehicles and transport; (7) the basic sciences of technology; (8) chemical industry; (9) food and luxury food industry; (10) industrial graphics (i.e. printing, photography, etc.); (11) textile industry; (12) clothing industry; (13) gemstones and industrial earths; (14) civil engineering; (15) industrial hygiene; (16) industrial safety; and (17) fire fighting and life-saving. This classification was retained until 1930, when there was a reorganization. Separate sections were created for the paper industry, for metrology and surveying and for music.

The outbreak of World War I in 1914 naturally had considerable consequences for the Museum of Technology, despite which work continued. In fact, essential parts were installed during wartime:

1914 *Ground floor*: railway museum, engine hall, 'lines of evolution' and various historical workshops (scythe forge, engraving and silversmith shop). *First floor*: hat workshop, tobacco industry and elements of the industrial graphics industry. *Second floor*: library, post museum, industrial hygiene.

1915 *Basement*: coal mine. *Ground floor*: shipping, agriculture, wood transport, river and torrent training works, mining industry and metallurgical engineering, a model of the Styrian iron ore mountain, a model of a mechanized dairy. *First floor*: clothing industry, chemical industry, basic sciences.

1916 *Ground floor*: completion of the engine hall, gas technology, history of lighting. *First floor*: surveying, horology, gems and rare earths, basketry, paper industry. *Second floor*: fire-fighting and life-saving.

1917 *Ground floor*: road vehicles, electrical engineering. *First floor*: civil engineering, textile industry, printing, Physikalisches Kabinett. *Second floor*: water supply, bridge construction, sewage engineering, historical aircraft (Lilienthalgleiter and Etrichtaube in the central hall).

1918 *Ground floor*: Styrian smelting hearth. *First floor*: chemical industry, room in memory of Karl Ludwig Freiherr von Reichenbach (1788–1869). *Second floor*: Cartography and geodesy.

1919 *First floor*: Large model of a sugar factory.

Other exhibition areas, such as the music section, were not, however, finished before the 1930s.[30]

There were good reasons for opening the museum before construction was completed. In particular, Exner, who was aware of the military situation and foresaw bad times ahead, pushed for an immediate inauguration, and thereby institutionalization of the museum, despite the fact that director Erhard wanted to wait until the whole building was finished.[31] Exner got his way. The museum was opened without any celebrations on 6 May 1918, the sobriety of the occasion matching the seriousness of the times (Figure 9).

Exner's decision was a correct one for the museum. It was now a functioning entity. Its further development, however, was strongly affected by the outcome of the 1914–18 war. The former Habsburg empire of some 60 million people was reduced to a rump state of German Austria, the so-called tadpole republic, containing just one-tenth of the former population. The most important industrial regions of the old empire

Figure 9 One of the first guided tours.

now lay in Czechoslovakia, while the bulk of agricultural land lay in Hungary. A victim of the shrunken economy and the autarkic policies of the successor states, the private Verein für das Technische Museum für Industrie und Gewerbe vanished. The directors of the Verein accordingly suggested that the museum be taken over by the Ministry of Commerce and Trade. In fact this act of nationalization took place at the beginning of 1922.[32] The 35 employees of the museum were taken on the strength of the Ministry. The Railway and Post Museum remained within the museum but retained its own administrative being.

The equipping of the museum carried through by director Erhard and his team followed Exner's plans designed to show the evolutionary history of various objects. So within the area of prime movers, for example, the development of the steam engine was shown, as was the evolution of spinning machines, and, among machine tools, the evolution of the lathe. Entire processes of production were also shown, as, for example, in the representation of the tobacco industry. Furthermore, Erhard made some effort to demonstrate essential stages of development by graphical means. An example may be seen in Figure 10, which shows water-lifting devices.[33] In general, the developmental stages were to be shown as (a) the stage of organic materials and artefacts; (b) the transition from organic to inorganic materials; and (c) the age of inorganic materials and artefacts. Such a triple division corresponded not only to the materials but also to the energy sources involved. This was to follow an idea Erhard had first developed in an address given in 1918:

the museum's function was to point out 'how the technology of economic labour has developed from the initial use of organic materials and human and animal energy to a situation in which inorganic nature is brought under control, and new sources of power and energy likewise'.[34] A late example of this developmental style of representation was the section on data processing planned by Professor Zemanek in 1974.

The exhibited objects were displayed in a museum designed in the prevailing style of the times, that is, the Austrian variety of art nouveau known as the *Jugendstil*. The wooden showcases were permanent fixtures but pleasing in overall appearance. This consistency of appearance was sometimes violated in work carried out after 1945. Of course, when the museum re-opens, moveable showcases will be absolutely essential. Interactive hands-on objects were of minor importance in the old museum but in each section at least one exhibit, if not a series of exhibits – sometimes engines, sometimes models, sometimes pieces of apparatus – could be set in motion. So, for instance, a sectioned locomotive, a model of a dredger from 1839, several kinds of steam engines (including one large one in the Gothic style), a paper-making machine, a blowing engine from a blast furnace of 1848 and several mechanical musical instruments were among the exhibits able to be set in motion at set times.

PUBLIC ACTIVITIES AND NUMBERS OF VISITORS

Apart from the permanent exhibitions the museum also held temporary exhibitions to increase its public appeal. The first of these special exhibitions was held in 1929 and was entitled 'The Economics of Water Power in Austria'. Another exhibition held in that year was entitled 'Austrian Coal', this theme being particularly apposite since Austria had lost most of its major coalfields on the break-up of the empire.[35] Other activities were the very popular slide lectures and film shows held on Saturdays and Sundays, the assembly hall with its cinematic equipment being well suited for such lectures and shows. Furthermore, between 1919 and 1922 there appeared a monthly periodical entitled *Technik und Kultur*, edited jointly by the Museum of Technology and the Freie Vereinigung für Volksbildung (the Independent Association for Public Education). Papers were contributed by Exner, Erhard and other curators of the museum, by professors from the Technical University, and writers such as Max Eyth. The periodical ceased to appear in 1922 because of financial constraints.[36]

The public's response to the range of activities by the museum was enthusiastic. Within the first six weeks of opening, 21,000 people had visited the museum, despite the hunger and cold that the citizenry were experiencing at that time. By 9 March 1919, the museum had welcomed a cumulative total of 100,000 visitors. During 1921 as many as 167,740 people visited the museum, a figure that was scarcely exceeded after 1945. Between 1918 and 1990 the museum received in total some eight million visitors.

Figure 10 Graphic concerning lines of evolution.

THE RESEARCH INSTITUTE FOR THE HISTORY OF TECHNOLOGY

As has been mentioned already, Wilhelm Exner was from the outset interested not only in technology but also in its history. For this reason he wanted more than mere displays: he wanted to show things in their technical and historical context. A modern museum of technology had, in his view, also to comprise:

1. A research laboratory for testing of materials and also for restoration and conservation.[37] This laboratory, although installed in the museum and provided with equipment, was never actually in operation by reason of lack of funds and personnel. It was finally wound up when the museum was shut down in 1992.
2. A research institute for the history of technology. This became a reality in 1931.

Unfortunately, Exner died before the research institute came into being. An earlier attempt to create such an institute had broken down at the time of the break-up of the empire.[38] Exner did, in 1928, attempt to set up a chair in the History of Technology at the Technical University of Vienna. Despite support from the Niederösterreichische Gewerbeverein (Association of Lower Austrian Manufacturers), the suggestion was turned down. The professors of the university took the view that the history of technology could be parcelled out among the particular faculties. They were not hostile, however, to the study of the history of technology as such, and suggested that a research institute be set up. The congress of the Austrian associations of engineers and architects also approved the idea and felt that the institute should be housed within the Museum of Technology. At this time the extension of protection to technical monuments was strongly urged.[39]

Even as late as 1929, the year in which his *Erlebnisse* (Memoirs) appeared,[40] Exner was still complaining that his repeated requests that the museum's treasures be made the subject of adequate scientific study and research, so important for the history of technology, had been consistently ignored. The seed, however, had been sown. Following reports by the Ministry of Education and the Academy of Sciences, the Ministry of Commerce and Traffic formally requested Erhard, as director of the Museum of Technology, to prepare a report on these matters. As a result, Erhard wrote a memorandum dealing with the concept of a Forschungsinstitut für Technikgeschichte (Research Institute for the History of Technology), the necessity for such, its tasks, the cost implications and the potential means of financing it. This memorandum was sent to a wide range of interested institutions.[41]

In March 1930 the Niederösterreichische Gewerbeverein took up the question of a research institute and set out 30 categories of industrial activity which should be properly researched. They also asked for a proper inventory of the Museum of Technology's collections to be prepared. In honour of Exner's ninetieth birthday, Oskar Miller, the founder of the Deutsches Museum in Munich, gave a lecture on technical monuments at the Gewerbeverein, giving further publicity to the idea of a research institute. A further lecture that helped to promote this idea was given by a leading historian of technology, Conrad Matschos, director of the VDI (Verein Deutsches Ingenieure; Association of German Engineers) at its congress in Vienna. On the occasion of this congress, Ludwig Erhard opened a special exhibition entitled 'Austrian Technology in Contemporary Documents' held in the Albertina – which held the collection of graphic materials – at the Museum of Technology and Industry. The Minister of Education at the time, the historian Heinrich von Srbik, opened the exhibition with an address entitled 'The Connection of Technology with Culture' (*Die Kulturverbundenheit der Technik*),[42] emphasizing the importance of the subject. This was able support for Erhard's own efforts to bridge the gap between history and technology and to promote the proposed institute. At a reception for the VDI, President Miklas of Austria announced to the assembled Austrian and German engineers that the government would set aside money in the budget of 1931 for the setting up of the research institute. This meant that the Museum of Technology could now actively engage itself in the business and set up a steering committee to oversee the matter. The committee's first meeting took place on 7 March 1931 under the chairmanship of Exner. Shortly afterwards, on 24 May, Exner died. It would have been some comfort to Exner to have known that at the second meeting in June a management committee for the new institute was elected. The President was to be Dr Karl Holey, professor at the Technical University and *Generalkonservator* of technical monuments. Erhard was appointed head of the institute. According to its constitution the institute was to be located in the Museum of Technology and answerable to the Ministry of Commerce and Traffic. Its brief was to develop the following areas of study:

- to locate archives and documentation relating to important Austrian inventions and industrial undertakings;
- to locate Austrian museums and other collections having in their possession objects of technological importance;
- to document and establish the history of technical monuments;
- to develop experimental methods for the investigation of technical objects;
- to collect materials for the biographies of Austrian technologists and their surviving archives;
- to organize the compilation of an *Österreichische Biographie.*

To pursue these activities the following facilities were to be provided:

1. A catalogue of documents, biographies, technical objects in public and private museums and collections relating to Austria's industrial heritage.
2. A library for the history of technology completed by a collection of pictures and technical drawings.
3. Lectures and publications in the field of history of technology,in particular the setting up of a yearbook.[43]

The yearbook has appeared since 1932 under the title *Blätter für Geschichte der Technik*, later re-titled *Blätter für Technikgeschichte.* Ludwig Erhard acted as editor of this official publication of the Research Institute until his death in 1940. After this the institute was taken over by the state and subordinated to the director of the Museum of Technology, who now also became editor of the yearbook. The yearbook after 1948 began to carry book reviews, until 1961 commemorated anniversaries important in the history of Austrian technology, and from 1968 included the museum's annual report.

In order to supplement the work of the Research Institute an association, the Österreichisches Forschungsinstitut für Technikgeschichte (Austrian Research Institute for History of Technology), was set up in 1989. It organizes lectures and, since 1989, has been represented on the editorial committee of the yearbook.

ARCHIVES, LIBRARY AND INVENTORY OF COLLECTIONS

Various though the activities of the Research Institute for the History of Technology were, and are, some functions were not fully covered and were undertaken by other institutions. So, for example, the documentation of technical monuments and historical industrial sites is the special care of the academic staff at the Technical University dealing with art history and care of monuments (*Kunstgeschichte und Denkmalpflege*) and the state-run office for monuments (Bundesdenkmalamt). Even now there exists no catalogue of documents and pictures, etc., relating to the history of technology covering the whole of Austria. Progress has been made, however. A bibliography has been published, and to a large extent the archives of the Museum of Technology have been made available to the individual researcher. The archives themselves contain the following materials.

1. *Personal papers and records of firms*
   - (a) *Mechanical engineering*: Viktor Kaplan, inventor of the Kaplan turbine and professor at the Deutsche Technische Hochschule Brünn (Brno); Johann Radinger, professor at the Technische Hochschule Wien; Ferdinand Redtenbacher, founder of scientifically based mechanical engineering; Gustav Schmidt, professor at the Deutsche Technische Hochschule Prag.
   - (b) *Ship-building*: Joseph Ressel, inventor of a type of screw propellor; Joseph Ruston and John Andrews, British engineers; ship construction drawings, autographs and contracts of Boulton and Watt[44] relating to steamship navigation on the Danube.
   - (c) *Aircraft construction*: Ludwig Lohner; Franz Xaver Wels; Erich Meindl.
   - (d) *Automobile manufacture*: Jakob and Heinrich Lohner, coach, motor car and scooter manufacture; Gräf and Stift, car manufacture and pictorial documentation.
   - (e) *Railway and canal construction*: Aloys Negrelli, planner of the Suez Canal.

(f) *Electro-engineering, data processing, films*: Gustav Tauschek, pioneer of data processing; the ELIN Company, electro-engineering and pictorial documentation; August Musger, inventor of slow-motion photography.

(g) *Musical instruments*: Lauberger and Gloss, piano manufacture.

There are in addition various other bequests of archive material.

2. *Collection of autograph material.* The collection contains autographs of outstanding technicians and industrialists (for instance Joseph Ressel, Karl von Reichenbach, Hugo Altgraf Salm), further autographs relating to technical education, preservation of monuments, guild and craft regulations from the sixteenth century, decrees relating to building projects, specification of patents and mining reports.
3. Documents relating to the foundation of the Technical Museum. Papers relating to the fitting out of the museum for the display of its collections.
4. Papers relating to the history and architectural history of the museum.
5. Biographical papers: 300 files relating to various Austrian technologists, scientists and industrialists.
6. Archive of plans and drawings: these are classified according to the 30 (32) groupings of objects in the collections, and contain plans, drawings, architectural drawings (of the eighteenth, nineteenth and twentieth centuries) and photographs.
7. Artistic depictions of technical subjects, among which are 180 etchings by Erich Veit (some with printing plates).
8. Portrait collection: portraits in oil, prints from copper plates, lithographs, photographs and negatives, mainly of Austrian technologists and industrialists.
9. Collection of photographic negatives: about 70,000 negatives and slides.

The entire archive, including the collection of negatives, forms part of the museum library, which contains some 100,000 volumes. In the library holdings are many *libri rari* relating to the history of technology, the oldest among them being from 1547. Naturally the collection contains such standard works as Jakob Leupold's *Theatrum machinarum*, the *Encyclopédie ou dictionaire raisonné des sciences . . .* by Diderot and Dalembert, Johann Krunitz's *Ökonomische technologische Encyclopädie* and Johann Joseph Prechtl's *Technologische Encyclopädie.* The library holdings cover history of technology, history of the natural sciences, history of industry and trade as well as works of a general technical nature. There are in addition holdings of various periodicals, and collections of catalogues relating to world exhibitions and industrial brochures.

Last, but not least, the inventory of the collections, now being transferred to a database and so far covering 36,000 items, is an essential tool for research in the history. The collections and archives provide evidence not only of Austria's internal technological development but also of its technology's international affiliations and connections. The industrial policy of the Habsburgs during the eighteenth and nineteenth centuries led to numbers of British engineers moving to Austria and thus taking the latest British technology with them. Already, however, by the beginning of the nineteenth century, Austria was no longer simply a recipient of technology, and was becoming an exporter of important ideas to the rest of the world. A notable example here would be the idea of setting up polytechnic institutes specifically to teach technology, such as those set up in Prague and Vienna. From the middle of the nineteenth century Austrian manufactures were already able to compete in terms of quality in international markets. Of course it is no part of this paper's business simply to celebrate the objects in the collections of the Museum of Technology. Its real function is to document how in the course of technological development what one is really seeing is a constant interchange of impulses across the industrial world.

Notes and References

1. Viktor Schützenhofer, 'Vom k.k. Fabriksproduktenkabinett zum Wiener Technischen Museum von heute', *Blätter für Technikgeschichte*, 1947, 9: 3.
2. Cameralism is the German, in particular Austrian, way of mercantilism, with systematic promotion of economy and the foundation of state-owned plants.
3. Hofkammerarchiv Wien, Kommerz, N.Ö. Fasc. 149: 410, ex April 1772.
4. J. Slokar, *Geschichte der österreichischen Industrie und ihre Förderung unter Kaiser Franz I* (Vienna, 1914), 54. V. Noback, *Über die erste Gewerbeausstellung anno 1791 nebst einer geschichtlichen Skizze der allgemeinen Gewerbeausstellungen in Böhmen* (Prague, 1873).
5. Viktor Schützenhofer, 'Aloys von Widmansstätten', *Blätter für Technikgeschichte*, 1947, 9: 34.
6. Ch. Hantschk, *Johann Joseph Prechtl und das Wiener Polytechnische Institut* (Vienna, 1988).
7. J. J. Prechtl, *Jahrbücher des kaiserlich königlichen polytechnischen Instituts in Wien*, 1819, 1: 25–33.
8. Prechtl, *op. cit.* (7), 27.
9. *Ibid.*, 68–9.
10. *Ibid.*, 67.
11. G. Maresch and H. Janetschek, *Werkzeuge der Biedermeierzeit. Die Werkzeugsammlung G. Altmütter*, Catalog of the Museum of Technology (Vienna, 1980). Gerhard Maresch, 'Werkzeuge aus der Biedermeierzeit', *Bl.f.T.G.*, H.41,42,43, 1979/80/81: 45–52.
12. G. Altmütter, *Beschreibung der Werkzeugsammlung des k.k. polytechnischen Institutes nebst einem Verzeichnisse der in demselben enthaltenen Stüke* (Vienna, 1847).
13. Bericht über die erste allgemeine österreichische Gewerbesproduktenausstellung im Jahre 1835, Wien.
14. Schützenhofer, *op. cit.* (1), 8.
15. Gerhard Maresch, 'Die Anfänge der Mechanisierung der Landwirtschaft in Österreich. Dargestellt an Hand der Sammlung von Modellen landwirtschaftlicher Geräte und Maschinen des Technischen Museums Wien', *Bl.f.T.G.*, 1984/85, H. 46/47: 7–38.
16. W. Exner, *Erlebnisse* (Vienna, 1929).
17. W. Exner, *Der Aussteller und die Ausstellungen* (Weimar, 1866).
18. Exner, *op. cit.* (16), 126. Maria Habacher, 'Das Technische Museum für Industrie und Gewerbe in Wien', *Bl.f.T.G.*, 1968, 30: 3.
19. W. Exner, *Das Technische Museum für Industrie und Gewerbe in Wien* (Vienna, 1908).
20. *Ibid.*, 90.
21. *Ibid.*, 128.
22. *Ibid.*
23. *Ibid.*, 139. Habacher, *op. cit.* (18), 1– 71.
24. Exner, *op. cit.* (19), Fig. 1. Erwin Zesch, 'Baugeschichte des Technischen Museums für Industrie und Gewerbe in Wien', *Bl.f.T.G.*, 1968, 30: 73–88.

25. *Ibid.*
26. Habacher, *op. cit.* (18), 32.
27. Karl Holey, 'Ludwig Erhard', *Bl.f.T.G.*, 1942, 8: 10. Exner, *op. cit.* (16), 127.
28. H. Burger, *Maschinenzeit Zeitmaschine. Technisches Museum Wien 1918–1988* (Vienna, 1991), 99.
29. Habacher, *op. cit.* (18), 46.
30. Burger, *op. cit.* (28), 99–101.
31. Exner, *op. cit.* (16), 128–9.
32. Habacher, *op. cit.* (18), 60–7.
33. Ludwig Erhard, 'Zur Entwicklungsgeschichte der Technik', *Bl.f.T.G.*, 1932, 1: 3–25.
34. Holey, *op. cit.* (27), 16.
35. Burger, *op. cit.* (28), 43, 44.
36. Burger, *op. cit.* (28), 35.
37. Exner, *op. cit.* (19), 81–6.
38. Christian Hantschk, 'Das Forschungsinstitut für Technikgeschichte am Technischen Museum Wien', *Österreich in Geschichte und Literatur*, 1989, 5: 292.
39. Holey, *op. cit.* (27), 19–25.
40. Exner, *op. cit.* (16), 130.
41. Ludwig Erhard, 'Das Österreichische Forschungsinstitut für Geschichte der Technik', *Bl.f.T.G.*, 1932, 1: 204–8.
42. Heinrich von Srbik, 'Die Kulturverbundenheit der Technik', *Bl.f.T.G.*, 1932, 1: 1.
43. Erhard, *op. cit.* (41), 204–8.
44. The Museum of Technology and Industry also possesses an original Boulton and Watt engine of 1790.

Cranks and Scholars

GRAHAM HOLLISTER-SHORT

In 1588 a sumptuous book containing 195 engravings of machines and devices was published in Paris. This was Agostino Ramelli's *Le diverse et artificiose machine.* This book is easily the finest work in the genre known as the machine books, and, on such evidence as there is, the one which ultimately reached the widest audience.[1] Although Ramelli was a military engineer, and included large numbers of pieces of military equipment and items likely to be useful on campaign among his machines, he did not neglect other heavy-duty engines such as wind and water mills for grinding grain and for water pumping. Given this preoccupation with the material needs of life, it is all the more surprising to come upon engraving number 188. What he shows us here (Figure 1) is a scholar hard at work, of course, but also obtaining a certain amount of exercise, *per accidens* as it were, by revolving a large wheel. This formidable piece of library furniture is able to hold up to eight books, and enables the reader seated before it to bring each into eyeshot by simply pulling on the rims of the drum. The point of it all, one supposes, is to facilitate the collating of variant readings and of establishing a text without having constantly to lift and carry cumbrous tomes. Now, thanks to the elaborate gearing, concealed within the machine, which controlled the attitude of the shelves, the innocent reader could be quite confident that the books would not fall on him or on his feet as the wheel was turned. This last point is particularly important if one takes at face value Ramelli's commendation of the wheel as helpful 'especially [to] those who are indisposed and tormented by gout'.

Obviously, each of Ramelli's machines, including this library wheel, or 'Texaurocycle', stood in some kind of relationship with the technology of the day in actual use. Mostly, these contexts have not been studied, but, as it happens, this is not true of number 188, which received a great deal of attention, although this was now some years ago. Bert Hall, who studied this device in 1970, threw much light not only on the pre-history of the gearing Ramelli used in his wheel but also on the pre-history of revolving bookcases in general.[2] I shall not rehearse here what has already been well done. I shall, however, first of all offer an example of a revolving bookcase much nearer in time and milieu to Ramelli than those cited hitherto, and secondly suggest a still wider context. This will indicate other and more remote antecedents, which (like the mitre worn by bishops) may be related to Christian fish symbolism.

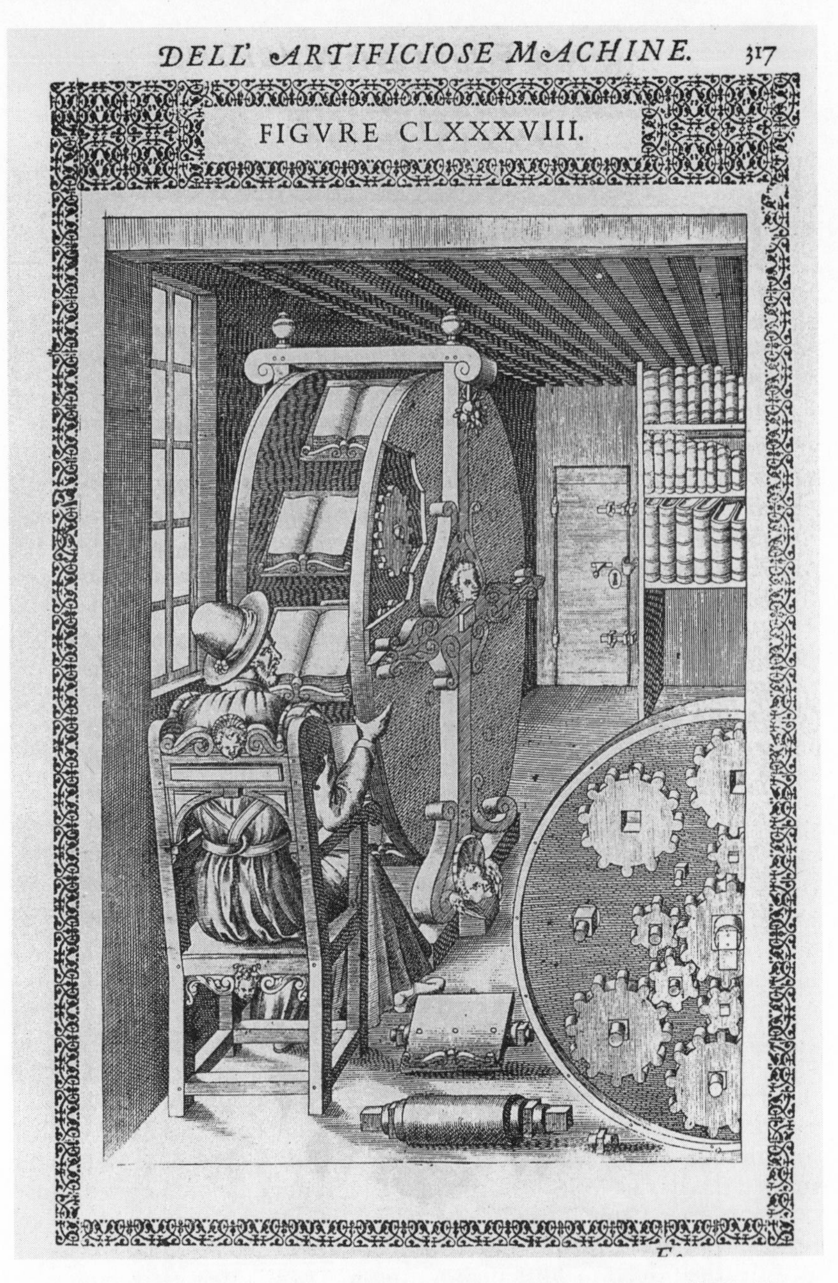

Figure 1 Agostino Ramelli, *Le diverse et artificiose machine* (Paris, 1588), Figure clxxxviii.

One of Machiavelli's *obiter dicta* was that *fortuna* and *virtù* ruled the world. Such a long sequence of accidental events, beginning with the decision of an antiquarian bookseller in San Francisco, had to happen before I could make this attempt, that I am fully persuaded that Machiavelli was right.[3] At all events, a Florentine publisher, Piero Pacini da Pescia, in 1514 chose as a frontispiece for his new edition of Bernardo de Granollach's *Ad inveniendum novam lunam* ... a woodcut showing a scholar in his study (Figure 2). It shows him, as befits the title of the book, calculating conjunctions and eclipses, for he is a student of that most noble art and science, astrology.[4] Beside him we see a square-bodied revolving bookstand mounted on a vertical axle which turned when the horizontal arm above the bookstand was given a push; unless of course it was really a lampholder. It looks as if four large volumes could be laid on it. The woodcut offers a clue as to what Ramelli may have been seeking to achieve in engraving number 188. Here the ordinary horizontal revolver has been rotated through 90° but by the application to it of *virtù* and epicyclic gearing is now able to hold double the number of books and defy the law of gravity while doing so.

Figure 2 Bernardo de Granollachs, *Lunare* ... (Florence, 1514), frontispiece.

The idea behind these devices, both simple and complex, that is, as an aid to scholarship, is appealing, and, as will appear presently, evidently filled a real need. It is easy to forget, with our modern, small, easily handled books, how very large and heavy medieval codices and Renaissance books could be. They were a problem to handle, especially perhaps for ageing scholars of no great muscular development. (It was for just such ageing scholars that spectacles had been invented some time before the beginning of the fourteenth century.) All the same, if this were all we knew of the matter, one might conclude that the Italian revolvers were simply a moderately interesting element in the material culture of southern Europe. But then, there was Czechoslovakia. What is to be seen there makes it quite clear that revolvers were not simply some sixteenth-century Italian fad, even if they marked new levels of sophistication in the business of book handling.

Before the Spanish conquered Mexico and Peru, and the mines of Zacatecas and Potosi began to flood Europe with silver, the mines of what is now central Slovakia supplied a large part of Europe's need for bullion. What were then called the seven 'Lower Hungarian mining towns' produced large quantities of gold, silver and copper. Pukanec, known in the German period as Buchants, and now merely a village, was one of these seven towns. It has a handsome church dedicated to Sv Mikulas (St Nicholas). A triptych over the altar (Figure 3), probably to be dated to *c.* 1450, has scenes in the side panels of particular interest. They show a pope, a bishop and a scholar (Figures 4, 5 and 6) all diligently pursuing their labours, their task lightened somewhat by the cranked lecterns at which they work. Obviously these devices are much simpler than the later Italian models but seem to perform the same function, that is, in making it an easy matter to swing a book within reading range, with the throw of the crank carrying it clear after one had finished.

The history of this idea can, however, be traced much further back, and, as will appear, may ultimately prove to have Byzantine roots. The Narodni Galerie on the Hradcany in Prague possesses three paintings figuring cranked lecterns or tables. One of these, by an unknown master of *c.* 1450, shows the Annunciation of the Virgin Mary. She is seated on a settle in front of what looks like a cranked dining-table (Figure 7). The two others, however, can both be dated to *c.* 1360. Both are from the hand of Master Theodorik, the greatest master of medieval Bohemia. They once formed part of a sequence of 127 panels, forming a pictorial wall, which he painted for the Chapel of the Holy Rood of Karlstein Castle, near Prague. This great work, in the tradition of Duccio's great altarpiece for the Cathedral of Siena (completed 1311), was almost certainly finished before the second consecration of the chapel on 9 February 1365. Since the first consecration took place on 27 March 1357, the painting of the panels in question probably falls between these dates. In a document of 1368, Karl IV, exempting Theodorik from taxes, speaks of him as '*pictor noster, praedictam capellam tam ingeniose et artificialiter decoravit*'.[5] One of Theodorik's two panels shows St Augustine working at a simple cranked lectern (Figure 8). The other, however, depicting St

Jerome, shows the great bibliophile and translator of the Gospels standing in front of a similar device, which, perhaps to show his superior standing, is fitted with a compound crank, rather poorly understood it has to be said (Figure 9).

But were revolvers an Italian fad and cranked lecterns a trans-Alpine curiosity? Not so, because it is clear that cranked lecterns were constructions to be observed on the fourteenth-century English literary scene. A group of five manuscripts were executed in about 1370, probably in East Anglia, for members of the Bohun family.[6] Among them is a psalter executed for Humphrey de Bohun, seventh Earl of Hereford, who died in 1373 (Figure 10). On f38r an illuminated letter E contains a picture of a saint reading from a cranked lectern. After I presented an earlier version of this paper in 1993, my attention was drawn to a work on miniatures from Armenia, some of which seemed to depict articles very like cranked bookstands, often supported on bases fashioned in the form of fish. More interestingly still, perhaps, and possibly as early as 1300, is a cranked lectern, also with a fish (Figure 11) from the Byzantine world.[7] It might well be, therefore, that both Eastern and Western Europe were drawing on some very much older tradition intimately connected with the Christian Church.

Figure 3 Sv Mikulas, Pukanec, Slovakia, triptych, *c.* 1450.

There remains something to be said about St Jerome's compound crank mentioned above. As things stand, this is the second oldest representation of the idea. The earliest occurs in Guido da Vigevano's *Texaurus regis Francie* . . . of 1335.[8] Guido, court physician to Philip VI of France, wrote his treatise for the king after the latter had taken his

Figure 4 Sv Mikulas, Pukanec, Slovakia, triptych, detail.

crusader's vow. The Christians were short of men. Guido's idea was that machines might help to make good this deficiency in recovering the Holy

Figure 5 Sv Mikulas, Pukanec, Slovakia, triptych, detail.

Land. One such device was a fighting car driven by a windmill (Figure 12). When the wind failed, the crew could propel the vehicle forward by

Figure 6 Sv Mikulas, Pukanec, Slovakia, triptych, detail.

applying themselves to the compound crank on the front axle. Guido's machine was the purest fiction but the windmill and the compound crank were not. The sails of the windmill, mounted on a mobile cap, plainly owe everything to the idea of the tower windmill, the earliest known record of which dates from 1294–5. Guido had taken the idea of the compound crank from the real world as well but what application it had there is lacking in contemporary documentation. There are two possible candidates, the carpenter's bit and brace, and the cranked

Figure 7 Unknown master, Annunciation, *c.* 1450. Courtesy of the Narodni Galeri, Prague.

lectern. I prefer the latter, if only because it would have been of large size; and, of course, Guido, being a scholar rather than a carpenter, might well have been accustomed to working at one himself.

Figure 8 Master Theodorik, St Augustine, *c.* 1360. Courtesy of the Narodni Galeri, Prague.

Notes and References

1. No one has attempted to gather references to Ramelli's work. Some are to be found in A. Keller, 'Renaissance Theatres of Machines', *Technology and Culture*, 1978, 19 (3): 507–8. To these may be added the very sour remarks on Ramelli's machines passed by Andreas Jungenickel, *Clavis machinarum* (Nuremberg, 1661), 4, and by Balthazar Rössler, *Speculum Metallurgiae Politissimum oder: Hell-polierter Bergbau-Spiegel* (Dresden, 1700), xx.

2. On revolving bookcases in East and West see J. Needham, *Science and Civilisation in China* (Cambridge, 1965), Vol. 4, Part II, 546–55, and B.S. Hall, 'A Revolving Bookcase by Agostino Ramelli', *Technology and Culture*, 1970, 3: 389–400, especially 390.

3. N. Machiavelli, *The Prince* (London, 1971), 130–1. The bookseller, in a sense the only begetter of this paper, was (and I hope is) Jeremy Norman, 442 Post Street, San Francisco. But his work would have availed me nothing but for Rupert Hall, who passed on the information, well knowing my taste for machination.

Figure 9 Master Theodorik, St Jerome, *c.* 1360. Courtesy of the Narodni Galeri, Prague.

4. The earliest edition of the many copies of Granollach's book in the British Library is one of 1485. This and all the others, however, lack the frontispiece so thoughtfully provided by Piero for the Florence edition of 1514.

5. U. Thieme and F. Becker, *Allgemeines Lexikon der bildenden Künstler von der Antike bis zum Gegenwart* (Leipzig, 1978), Vol. 32, 595–6. See also K. Stejskal, *Mistr Theodorik* (Prague, 1978), 14–17.

6. M.R. James and E.G. Millar, *The Bohun MSS. A Group of Five Manuscripts Executed in England about 1370 for Members of the Bohun Family* (The Roxburghe Club, Oxford, 1936). The name Himfridus occurs more than once in the psalter referred to in this paper. The manuscripts are considered stylistically to belong to the East Anglian school. The authors refer to the cranked lectern as 'a desk with right-angled bend in the shaft'.

7. A. Guevorkian, *The Crafts and Mode of Life in Armenian Miniatures* (Yerevan, 1973), plate 27, 7, and plate 28, 6. G. Cavallo, ed., *Libri e lettore nel mondo Bizantino. Guida storici e critica* (Rome, Bari, 1982), tav. 20. I have to thank Francis Maddison, Curator of the Museum of the History of Science, Oxford, for both these citations, and indeed for much additional help.

8. A.R. Hall, 'Guido's Texaurus, 1335', in *On Pre-modern Technology and Science. Studies in Honor of Lynn White Jr*, ed. B. Hall and D. West (Malibu, 1976), 11–52.

Figure 10 Duke Humphrey's psalter, 1371, MS 1826, f 38r. Courtesy of the Österreichische Nationalbibliothek, Vienna.

Figure 11 Portrait of Niketas Choniates, working at a cranked lectern, *c.* 1350. Cod. hist. gr. 53. Courtesy of the Österreichische Nationalbibliothek, Vienna.

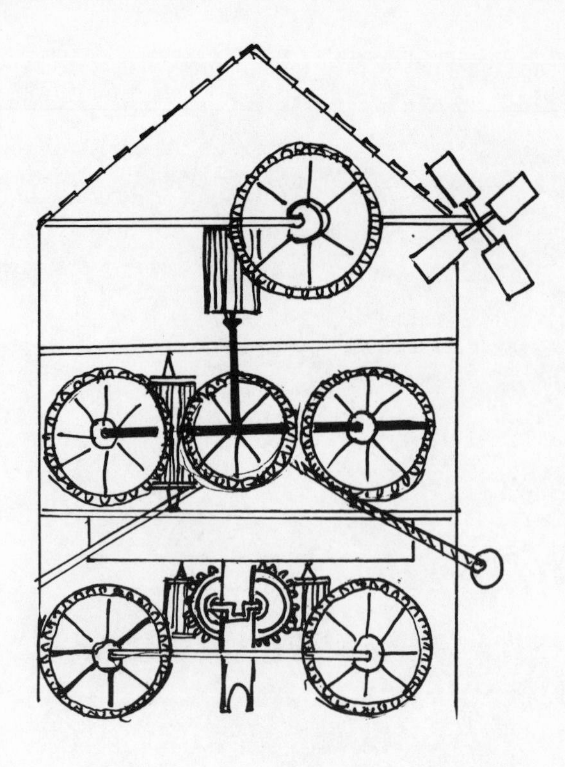

Figure 12 'On the way of making a . . . waggon which is propelled by the wind . . . ' After A. R. Hall, *op. cit.* (8), ch. xii. Note the cranked front wheels in case the wind fails.

Contents of Former Volumes

FIRST ANNUAL VOLUME, 1976*

SECOND ANNUAL VOLUME, 1977*

SEVENTH ANNUAL VOLUME, 1982*

EIGHTH ANNUAL VOLUME, 1983*

NINTH ANNUAL VOLUME, 1984*

TENTH ANNUAL VOLUME, 1985*

ELEVENTH ANNUAL VOLUME, 1986*

TWELFTH ANNUAL VOLUME, 1990

THIRTEENTH ANNUAL VOLUME, 1991

FOURTEENTH ANNUAL VOLUME, 1992

FIFTEENTH ANNUAL VOLUME, 1993

SIXTEENTH ANNUAL VOLUME, 1994

* Out of print.